Aquinas, Education and the East

Sophia Studies in Cross-cultural Philosophy of Traditions and Cultures

The Sophia Studies in Cross-cultural Philosophy of Traditions and Cultures fosters critical and constructive engagement of the intellectual and philosophical dimensions – broadly construed – of religious and cultural traditions around the globe. The series invites innovative scholarship, including feminist, postmodern, and postcolonial approaches.

For further volumes:
http://www.springer.com/series/8880

Thomas Brian Mooney • Mark Nowacki
Editors

Aquinas, Education and the East

Springer

Editors
Thomas Brian Mooney
School of Social Sciences
Singapore Management University
Singapore, Singapore

Mark Nowacki
School of Social Sciences
Singapore Management University
Singapore, Singapore

ISBN 978-94-007-5260-3 ISBN 978-94-007-5261-0 (eBook)
DOI 10.1007/978-94-007-5261-0
Springer Dordrecht Heidelberg New York London

Library of Congress Control Number: 2013931262

Printed on acid-free paper

Springer is part of Springer Science+Business Media (www.springer.com)

Foreword

It has been a commonplace of various fashionable intellectual movements influenced by hermeneutics, linguistic theory, and anthropology to insist upon the relativity of concepts, values, and modes of thought, even to the point of arguing that intercultural comparison is impossible. Set in sharp opposition to this, however, is the thought that if we are to regard the behaviour of others as expressive of beliefs and values then *ipso facto* we must be able to make sense of it, not simply in the respect of being able to translate it, but to the extent of seeing it as intelligible and even reasonable, which is to say reasonable by reference to a common set of beliefs and values.

If the first, *relativist* idea seems extravagant and disregarding of the extensive practices of translation and comparative studies, the second, *transcendentalist* one seems unduly rationalistic. In fact both are somewhat a prioristic, announcing antecedent to empirical enquiry what cannot or must be the case. Leaving to one side general philosophical presuppositions about the possibilities of interpretation, and considering what we know about the diversity of human societies, we might expect to encounter both strangeness and commonality, confirming that while human forms of life may differ in time and place, insofar as they are forms of human life they are also expressions of a common nature.

That very thought may then be looked for among the diverse human cultures and traditions, and sure enough it is to be found in one form or another in reflective writings from the ancient world in both Western and Eastern societies. In the *Histories*, for example, Herodotus provides descriptions of the beliefs, legends, and moral codes of various peoples defeated in Persian conquests and of others from further afield. Often his examples are meant to strike his readers as absurd or shocking, but on reflection several points emerge. First, Herodotus uses the fact that such diverse beliefs are strongly held by some and also firmly rejected by others, to suggest the formative and enduring influence of *custom*. He writes: "If anyone, no matter who, were given the opportunity of choosing from amongst all the nations in the world the set of beliefs which he thought best, he would inevitably—after careful consideration of their relative merits—choose that of his own country" (*Histories*, III).

Second, however, the point is made in terms that presuppose the possibility of close comparison. Third, differences are chosen in relation to a common theme, as when he reports Darius inviting Greeks and Indians to consider one another's treatment of the dead (burning and eating, respectively). The universality of custom, the possibility of considered comparison, and the commonality of underlying interests all point to the idea of a common human nature. This idea runs through the history of ideas from antiquity into recent centuries and continues to be presupposed even when it is apparently rejected, such is the pragmatically self-refuting character of universal relativism.

Moving from general assumptions of a common human nature to the content of particular conceptions of that nature, certain elements are central and recurrent within the range of traditions; for example, identifications of intellect, desire, and will; of reason and sentiment; and of self-regarding and other-regarding dispositions. Two millennia before Adam Smith in his *Theory of Moral Sentiments* built upon Hume's account of the inverse relation between the strength of moral feeling and the distance of its objects. Mencius observed that while the primary and proper focus of benevolence is upon one's parents "a benevolent man extends his concern from those he loves to those he does not love" (*Mencius* IV, B. 1.). Similarly, the Kantian idea that a condition of right conduct is that it should pass the test of universalisability is anticipated in the *Analects* of Confucius, as well as in Greek philosophical and near-Eastern religious and legal traditions.

Central also to the moralities of major cultures is the idea of virtue, that is, of dispositions of thought and feeling expressed in actions and reactions. An interesting and regrettable consequence of restricting discussions of historical treatments of virtue to those originating in Greek ethics is the relative brevity of the list, principally, of course, the cardinal four: justice, prudence, temperance, and fortitude. Reading through the *Analects*, by contrast, I count at least 70 virtues identified for respect and emulation, including attentiveness, circumspection, decency, empathy, determination, dignity, discernment, filiality, generosity, graciousness, hopefulness, humour, loyalty, moderation, modesty, patience, piety, resolution, refinement, resourcefulness, sagacity, simplicity, sincerity, and trustworthiness.

Of course, these dispositions have been identified via English translations of Chinese characters—in this case the translations of Edward Soothill, an English Victorian-cum-Edwardian missionary, educationalist, and Professor of Chinese at Oxford. In the process, Western and specifically Imperial British attitudes will have had some influence, and at one point, Soothill quotes Confucius as saying that "A wise man in regard to what he does not understand, maintains an attitude of reserve. If terms be incorrect, then statements do not accord with facts" (*Analects* VII, III 4–5). As it happens, however, Soothill was a great admirer of what he recognized to be a code of gentlemanliness in traditional Chinese culture not so very different from that prevailing in his own society, and this opened him to Confucius's fine-grained classification of cognitive, affective, and executive virtues.

The Western code that informed Soothill's reading was itself developed in part out of medieval chivalry traditions which enjoyed a marked revival during the reign of Queen Victoria, and were consciously introduced into reflections on the aims of education. In their original and revived forms, these traditions sought to embody the four cardinal and three theological virtues (faith, hope, and charity), but alongside them ran other traits that were thought to be natural accompaniments, either as expressions of or as aids to the principal virtues.

In this connection it is interesting to read Thomas Aquinas' short letter to Brother John *De modo studendi*—'On how to study', and to note the points of resemblance to Confucian instruction. Aquinas writes as follows:

> Since you asked me, my dearest in Christ Brother John, how you should study in order to acquire the treasure of knowledge, I offer you this advice on the matter: Do not wish to jump immediately from the streams to the sea, because one has to go through easier things to the more difficult. Therefore the following points are my warning and your instruction. I command you to be slow to speak, and slow to go to the conversation room. Embrace purity of conscience. Do not give up spending time in prayer. Love spending much time in your cell, if you want to be led into the wine cellar. Show yourself amiable to all. Do not query at all what others are doing. Do not be very familiar with anyone, because familiarity breeds contempt, and provides matter for distracting you from study. Do not get involved at all in the discussions and affairs of lay people. Avoid conversations about all any and every matter. Do not fail to imitate the example of good and holy men. Do not consider who the person is you are listening to, but whatever good he says commit to memory. Whatever you are doing and hearing try to understand. Resolve doubts, and put whatever you can in the storeroom of your mind, like someone wanting to fill a container. Do not spend time on things beyond your grasp. Following such a path, you will bring forth flowers and produce useful fruit for the vineyard of the Lord of Power and Might, as long as you live. If you follow this, you can reach what you desire.

Confucius, meanwhile, is credited with the following:

> The scholar who in his food does not seek the gratification of his appetite, nor in his dwelling is solicitous of comfort, who is diligent in his work, and guarded in his speech, who associates with the high-principled, and thereby directs himself aright, such a one may really be said to love learning (I, xiv). …
>
> He who knows the truth is not equal to him who loves it, and he who loves it is not equal to him who delights in it (VI, xviii). …
>
> The scholar who becomes widely versed in letters and who restrains his learning within the bounds of good conduct is not likely to leave the track (VI, xxv). …
>
> Learn as if you were not reaching your goal, and as though you were afraid of missing it (VIII, xvii).
>
> I have spent the whole day without food and the whole night without sleep in order to think. It was of no use. It is better to learn (XV, xxx).

Separated by seven and a half centuries, five thousand miles, and the idea of monotheism, there are nonetheless significant points of resemblance between these reflections on the demeanour and aims appropriate to learning. From the point of view of philosophy of education the significant difference between the authors is that Confucius' aphorisms stand apart from any theoretical account of epistemology, metaphysics, and value theory, while these latter fields more naturally characterise the work of Aquinas than do words of wisdom. Among the reasons for this is that whereas Confucius was a sage, Aquinas was a philosopher-theologian.

This suggests a way of studying Aquinas and Eastern sage traditions as complementary, for the focus of the writing of Confucius and Mencius, as those of Lao Tzu and Zhu Xi, is upon the guidance of conduct through the cultivation and internalization of certain habits and ritual practices; whereas Aquinas gives his attention to the analysis of action and cognition and the identification of the modes of causality involved in these. The effort to read Confucius in light of an analysis of material, formal, efficient and final causation, or by reference to the distinction between innate and acquired dispositions, or in connection with an account of the good as pertaining to the realization of nature, is neither difficult nor unrewarding. It is not something, however, that he invites, and certainly his own methods are not theoretical.

Today, across the world but especially in the West, thinking about education moves uncertainly between different sets of aims. On the one hand talking about the importance of *personal growth* and *civic virtue*; on the other emphasizing the need to develop *future-oriented skill-sets* and *adaptability to a knowledge-economy*. A natural response to such stated aims is to doubt that they have much in the way of real content, noting that they leave unaddressed, let alone resolved, the more fundamental issues of the modes and ends of growth, the substance and objects of virtue, the value of the purposes to which skills might be directed, and the relation between the intrinsic and instrumental value of knowledge. In the East meanwhile there is a corresponding contrast between the traditional modes of formation directed towards cultivating fittingness of demeanour in the face of the contingencies of the human condition, and the development of skills adapted to technological and economic progress.

The way in which I have introduced these different aims may suggest acceptance of some of the familiar ways of opposing them, but in fact it seems an open question as to whether they need be in tension. Resolving that question is in part a task for the social sciences but at a more fundamental level it is also a challenge for philosophy of education. What Aquinas shows very clearly is that while practical questions call for practical answers they also raise, through a series of further questions related to the successive ends implied in chains of practical reasoning (why do this? so as to get those; why aim for those? in order to attain that? etc.), issues about the nature and value of various goals and their relationship to the agents who pursue them. These belong to the metaphysics of value, and to other parts of theoretical philosophy, and progress in them is impossible without engaging with notions of truth and objectivity, substance and causality, and agency and intention.

It is to the study of these notions that Aquinas has most to contribute to the philosophy of education, and to the integration of systematic enquiry and sapiential guidance, and thereby to a dialogue between the speculative orientation of Western philosophical thought and the sage traditions of Eastern cultures. Here I leave to one side the further fields of theology: scriptural, systematic, and spiritual, as these have developed in Jewish, Christian, Islamic, Hindu, Buddhist, Sikh, and Jainist thought—not as irrelevant, but as engaging a whole further dimension (in which, of course, Aquinas was again a giant).

Brian Mooney and Mark Nowacki have been active and also effective in nurturing such enquiries. First, in conceiving and convening the conference in which many of the following diverse and very interesting chapters originated. Second, as scholars on their own account studying, in some part translating, and providing insightful commentaries on three major figures in the Western Christian tradition of reflection on the nature and ends of education: Augustine (*De Magistro*); Newman (*The Idea of a University*) and again Aquinas (*De Magistro*). Armed with this present volume on *Aquinas, Education and the East*, and with Mooney and Nowacki's *Understanding Teaching and Learning: Classic Texts on Education* (Exeter: Imprint Academic, 2011), we are well placed to engage the scholarly and practical issues involved in bringing Western and Eastern thought on education into fruitful exchange. Participating in the original conference, then reading these chapters, I sense the beginnings of future work that could have real practical as well as intellectual benefits.

University of St Andrews John Haldane

Acknowledgments

This book is the culmination of a process that began in 2009 when the Editors of this volume organised a symposium entitled 'Aquinas, Education, and the East', at Singapore Management University. That symposium was generously funded by the Wee Kim Wee Centre at Singapore Management University and the Editors are most grateful for the Centre's support and, in particular, to its Director Kirpal Singh. We are also grateful to His Grace Nicholas Chia, Archbishop of Singapore, who opened the symposium.

As with many symposia and conferences, some of the most interesting and exciting developments occur 'outside' the official proceedings. John Haldane was the keynote speaker at the symposium and it is largely due to his estimation of a selected number of the papers presented and his encouragement that we chose some of the best papers and requested that they be reworked for publication. We also solicited a number of papers from scholars who were unable to attend the conference. In this we are indebted to Eleanore Stump who was genuinely enthusiastic about the volume and who helped us to identify a number of scholars working in the neglected areas that represent the themes of this book. Throughout the sometimes difficult editorial process, we have been buoyed by the support of Alasdair MacIntyre and the late Father James McEvoy. We would also like to thank Steven Burik for his assistance with the preparation of the manuscript.

In addition, the Editors would like to thank our families Pat, Fionnuala, Thomas, Chuen, Dodge, Tippy, Callie, and Plato.

Contents

Part III Education and the East: Reflections on Educational Policy

Introduction

T. Brian Mooney and Mark Nowacki

The influence of St Thomas Aquinas' theological and philosophical thought is immense; few have had as profound an impact on the theology and practice of Western Christianity in the past 1,000 years, and he occupies a secure place in the philosophical pantheon with Plato, Aristotle, and Descartes. Nonetheless, significant dimensions of Aquinas' intellectual achievements have yet to receive due consideration. The present volume contributes to several under-explored areas in contemporary Thomistic scholarship, and takes as its focus the relation between Aquinas and Eastern thought, specifically in regard to education. What can Aquinas teach us about the theory and practice of education? How do his views relate to the great traditions and thinkers of the East? This volume of new essays is thus an overture (and hopefully a catalyst) to further reflection and studies on these themes.

Aquinas has much to say on the philosophy and practice of education, but he nowhere in his vast corpus devotes himself to a systematic treatment of the field. The most extended discussion of a philosophical and theological approach to education occurs in Question 11 of the *Disputed Questions on Truth*, commonly referred to as the *De Magistro* (*On the Teacher*). Much crucial context for the *De Magistro* is provided in the *Summa Theologica*, but relevant discussions are scattered throughout his works.

In the *Summa Theologica*, Aquinas takes up the Augustinian (and Neoplatonic) image of education as a circle of enlightenment. (*ST* I.89.1) God is the source of all enlightenment, truth, and knowledge. God radiates out 'an intellectual light' that suffuses the entire created order. Understanding is assimilated or attenuated in relation to where a being is situated within the ontological hierarchy of created beings. Creatures ontologically closer to God participate more fully in this intellectual light

T.B. Mooney (✉) • M. Nowacki
Singapore Management University, Singapore, Singapore
e-mail: brianmooney@smu.edu.sg; nowacki@smu.edu.sg

T.B. Mooney and M. Nowacki (eds.), *Aquinas, Education and the East*,
Sophia Studies in Cross-cultural Philosophy of Traditions and Cultures 4,
DOI 10.1007/978-94-007-5261-0_1, © Springer Science+Business Media Dordrecht 2013

and creatures at a further remove participate to a lesser extent. By implication, and given its relative proximity to God, an angel would participate more fully and interiorly in this intellectual light than would a human being.

It is this background thematic that permeates the seemingly odd question that opens the *De Magistro*: 'Can a human teach and be called a teacher, or just God alone?' For Aquinas, since God is the *fons et origo* of being and understanding, there is, for him, a deep sense in which all teaching, learning, and understanding derive from God, and God is our only true teacher. Aquinas, of course, admits a role for human teaching, but all human teachers teach only in an approximate and imperfect manner. No matter how gifted the human teacher may be, the pedagogical process requires acts of willing receptivity and interior judgement by the student. Throughout, God is a dynamic contributor to the process of education, simultaneously enlightening teacher, student, and what is taught.

These themes, which are also reflected in Eastern thinkers and traditions, are taken up in several essays in this volume. One rich vein for comparative work on Aquinas and Eastern traditions involves engagement with the Hindu and Buddhist understandings of *Jnana*, wisdom attained through meditation. In Hinduism, we encounter the notion that the highest forms of knowing involve a form of self-realisation, that *Atman* is *Brahman* and that *Brahman* is *Atman*, which parallels the Christian formula, that God became Man so that Man could become God. Or, in a Buddhist modality, we find that the highest form of awareness entails freeing oneself from conceptual encumbrances and divisions, which has theological echoes in the unity-within-diversity of the Christian Trinity. Moreover, for both Aquinas and Buddhists, attachment to the world is a source of suffering and division from our true selves. Conceptual elaboration of these tropes often prompts representatives of Eastern and Western traditions to draw upon similar conceptual resources and distinctions, as several of the contributors to this volume amply demonstrate.

At a practical level, and especially in regard to moral education, there are several points of contact between Aquinas and the great traditions and thinkers of the East. There is much agreement about what questions are important to ask, and there are startling commonalities and differences with respect to how those questions should be answered. How can morality be taught—if it can be taught at all? What is involved in moral knowing? How are virtues formed? Are there specific practices or perspectives that support or undermine virtue formation? How might we best compare and coordinate the views of Aquinas and profound thinkers of the East on understanding, teaching, and learning? How have (or perhaps should) Eastern and Thomistic insights be embodied in concrete educational practice? What implications either do or should these insights have, at an operational level, for education policy? These questions, in turn, are taken up by several of our authors.

To open out these questions in a more systematic fashion, the essays of this volume have been arranged into three Parts. In Part I, Aquinas and Education, our essayists take up the task of understanding and extending the thought of Thomas Aquinas. These essays provide necessary background for understanding the highly-nuanced view of St Thomas on education in general and on moral education in particular. In Part II, Aquinas and the East, our essayists adopt explicitly comparative approaches.

Aquinas' theological and philosophical thought is coordinated with a variety of significant thinkers within the Eastern and Western canons. Finally, in Part III, Education and the East, our essayists consider specific educational policies and proposals. Their handling of these proposals demonstrates, on the one hand, how Thomistic resources can be used to address current policy challenges, and on the other hand, how educational policy has been implemented within the context of an Eastern educational system with a Catholic heritage.

Our first essay, by Jānis (John) Ozoliņš, provides useful scholarly background and context for Aquinas' general accounts of understanding, teaching, and learning. Readers less familiar with the contours of St Thomas' thought would do well to start with this essay. Ozoliņš brings out how understanding the Divine role in education shapes what Aquinas counts as knowledge and what it means to say that someone has learnt. While his essay represents an insightful scholarly account of St Thomas' key themes on education, Ozoliņš is concerned throughout to connect up medieval concerns with their contemporary analogues in the practice and theory of education.

The second essay, co-authored by the Editors of this volume, aims to restore the traditional notion of connatural knowledge to its proper place in an acceptable account of teaching and learning. In the broadest sense, connatural knowledge is knowledge readily acquired by beings having a certain nature, much as dogs have a ready access to a world of scent much richer than that known to humans. Full possession of the virtues, we argue, involves connatural knowing. Connatural knowledge emerges as a knowledge by inclination which systematically tracks the specific moral interests humans possess precisely because they are human. This essay draws out the implications of a central Thomistic theme and provides a new angle on the connections among know-how, virtues, and skills. As a result, the essay provides a novel approach to themes in epistemology and, in particular, contributes to the fertile new field of virtue epistemology.

Andrew Pinsent's remarkable third essay charts what we consider to be an exciting new direction in virtue ethics. While it is true that Aquinas develops his ethics within the Aristotelian tradition, nonetheless St Thomas introduces several novel elements into his account of the moral life. Familiar Aristotelian virtues are interwoven with non-Aristotelian attributes: gifts, beatitudes and fruits. For instance, Aquinas holds that wisdom, properly understood, is a gift appended to the virtue of *caritas*, or divine friendship. Pinsent argues, following Aquinas and certain insights gathered from contemporary psychology, that the appropriate locus of virtue formation is not, as most would have it, at the level of the first person, but rather at the level of second-person relatedness. Pinsent educes the Thomistic insight that one's growth in the virtues requires a shared identification with the intentional stance of another person. He illustrates this with an analysis of the psychological phenomenon of joint attention and describes the manner in which it illuminates autism. The role of an educator can thus be seen to be inherently second-personal: the ancient understanding of the intimacy of the teacher/pupil relation then finds confirmation in contemporary science and, by implication, points to significant lacunae in contemporary educational theory and practice.

The fourth and final essay in Part I, by Thomas Ryan, brings Aquinas into dialogue with Elspeth Probyn, who has made significant contributions to the sociological study of shame. According to Ryan, the profile of shame elucidated by Probyn resonates well with the universalist tendencies of Thomistic ethics. Ryan hopes to gain a clearer picture of the educative and transformative function of shame in the personal, social, cultural and moral dimensions of human life. Given the ubiquitous normative force of shame in Asia, exploration of the experience of shame can help bridge ethical approaches found within Eastern and Western traditions.

Part II opens with an essay by Anh Tuan Nuyen, who advances the provocative suggestion that Aquinas' query 'Can a human teach and be called a teacher, or just God alone?' (*De Magistro*) should be understood as a question primarily about moral education. Interpreted thus, Aquinas could be seen to be providing an answer to the moral sceptic. This take on Aquinas allows for fecund comparisons with Confucian accounts of moral education, and in particular with the thought of Mencius. Further comparisons to Xunzi allow Nuyen to set an agenda for deeper and richer comparative speculation on how Eastern and Western traditions might be further articulated, particularly with respect to the cultivation of the self.

In the next essay, Doug Mikkelson brings to light some implications of the indisputable fact that Aquinas and Dōgen devote considerable thought to the moral education of 'beginners' in the religious life. A key strategy employed by both is that of proposing moral exemplars. The exemplars adduced range from central cases—Christ and Buddha—to expert, but less exalted, practitioners (saints?) in their respective traditions. Interestingly, the centrality of exemplars in moral education has all but disappeared from contemporary moral philosophy. Mikkelson's timely analysis reminds us of the value and importance of the sapiential dimension of philosophy, which formerly provided much of the justification for philosophy as a choice-worthy pursuit for human beings, and which motivated recognition of philosophy as a distinct way of life.

The metaphysical simplicity of the Ultimate has long been a central theme in Eastern and Western philosophical and religious traditions. St Thomas' commitment to God's absolute simplicity is one of, if not the central, metaphysical insight that drives his philosophical theology. This commitment creates immediate tensions in Aquinas' thought since he also maintains the orthodox Christian position that God is Triune. After exploring these dimensions of St Thomas' thought, Joseph O'Leary calls attention to and elucidates Eastern analogues of this problematic. The insight that the Ultimate must be simple is upheld against a backdrop of logical and experiential considerations that militate in favour of multiplicity.

Cecilia Wee, our final essayist in Part II, illuminates our understanding of Aquinas by comparing his theory of knowledge and philosophy of education with the views of John Locke. As Wee demonstrates, there are significant points of contact between the two broadly empiricist philosophers. She points to important commonalities between Aquinas' account of *scientia* and Locke's account of knowledge, and notes that both philosophers hold that this higher form of human knowing is to be distinguished from lesser epistemic states such as belief and opinion. Wee sketches an account of how accepting an empiricist theory of knowledge shapes one's theoretical understanding of learning and, in particular, impacts one's understanding of the role of the teacher in learning.

In Part III, our essayists take up a different perspective and roll up their sleeves to talk about educational policy. The first essay, by Jude Chua, takes up a suggestion by Francis Davis and Nathan Koblintz that Catholic schools in England and Wales should be reorganised into social enterprise zones. Chua extends this idea in two directions. First, he suggests that schools so organised be re-conceptualised as 'play schools', wherein stakeholders playfully participate in the project of education. Second, he argues that inculcation of a social enterprise dimension allows schools to become springboards for what he terms 'gifted education'; namely, an epistemological and moral awakening with respect to one's own normative biases. Chua's essay explicitly draws on Aquinas' metaphysics of Divine play, and teases out implications of the social embeddedness of the educational enterprise, thereby engendering a dialogue on ends and means among participants both internal and external to the school.

In the final essay, Andrew Crow and Thomas O'Donoghue reflect upon educational reform in the Philippines—a country with deep roots in Catholic social and education practices yet located within a distinctively Eastern context. As is appropriate given the policy focus of the paper, the authors provide an impressively detailed but concise historical review of Philippine educational practice and policy during the past two decades. The resultant narrative displays how policy makers concretely go about the difficult task of educational reform whilst trying to align public expectations, educational research, and principles of universal access against evolving global benchmarks and standards. Their essay does not address Aquinas at the surface level. Nonetheless, the construction of their narrative, and more particularly, the principles of selection they employ for what is included (and excluded), together with the underlying social principles taken as normative, conjointly display a deep engagement with the Thomistic educational tradition in concrete practice and ethos. This is an example of how policy makers actually argue and support their positions, and Crow and O'Donoghue are explicitly engaging policy makers in their own terms. The essay thus provides us with a blueprint for how philosophical principles can be translated into a mode accessible to policy makers and thereby effectively shape the implementation of educational initiatives.

Part I
Aquinas and Education: Understanding and Extending Aquinas

Aquinas and His Understanding of Teaching and Learning

Jānis Tālivaldis Ozoliņš

1 Introduction

The nature of the educative process continues to be hotly debated not just because the purposes of education remain contested but also because there is little agreement about how the efficacy of teaching can be improved so that students can learn more successfully. There is at least agreement on one point, namely, that teaching (at a minimum) has the aim of enabling pupils to learn what is worth knowing, whatever other functions it might have. Universities and other higher education providers prescribe models of teaching and learning, and demand that teachers within them adhere to the established teaching and learning paradigms.[1] Michael Peters writes that in most institutions, philosophy of teaching usually refers to a statement of an individual's teaching philosophy, which is generally a statement about teaching practice, rather than a statement about a philosophy of teaching (Peters 2009).[2] Nevertheless, despite the difficulties in articulating a clear statement about what philosophy of teaching an individual teacher adopts, it is important to have some conception of what it is that is going on in the educative process and some theoretical justification for the particular teaching decisions made. Every teacher needs to enter into some reflection on whether his or her essential approach is one which adopts a transmission model of teaching or a facilitation model of teaching or perhaps a combination of these two approaches.[3] Other substantive questions arise, especially in relation to these two broad models, concerning the nature of knowledge. If knowledge is innate, we might be inclined to think that all we need to do is draw it out of the student, much as Socrates famously does in Plato's *Meno* (Plato 2005). That is, the teacher's job is to facilitate remembering. On the other hand, if knowledge depends on experience and is hard won through observation and theory

J.T. Ozoliņš (✉)
Australian Catholic University, Melbourne, Victoria, Australia
e-mail: john.ozolins@acu.edu.au

T.B. Mooney and M. Nowacki (eds.), *Aquinas, Education and the East*, Sophia Studies in Cross-cultural Philosophy of Traditions and Cultures 4,
DOI 10.1007/978-94-007-5261-0_2, © Springer Science+Business Media Dordrecht 2013

construction, we might be inclined to adopt a transmission model, for we would want the next generation to benefit from what has been painstakingly accumulated.

In this paper, it is not our intention to address the broader issues that arise in relation to teaching and learning. Not only are there questions about whether we should adopt a transmission or facilitation model of teaching, but also whether we should adopt a traditional approach or a progressivist approach.[4] Still other questions concern the nature of knowledge itself. Our aim in this paper is modest. We will be concerned to elaborate, from various sources, Aquinas' view of teaching and learning. We would like to think that if he were to be asked to provide us with a teaching portfolio describing his approach to teaching, he would furnish an outline such as we develop below. For Aquinas, teaching is connected with the Divine, since he argues that though human beings are able to teach, they do so in a secondary sense and that it is God who primarily teaches. This is because God is the source of all being and is the light at the heart of our being. In the learning process, a key feature of Aquinas's account builds on the nature of illumination, which is to say an understanding of what is taught that enables us to see how what we have learnt connects to other things. Ultimately, these connections lead us to Wisdom, which is to say God, and for Aquinas wisdom in its different forms is the central aim of all teaching and learning.

2 Teaching and Learning

Although Aquinas does not develop a treatise on teaching and learning, he spent a considerable amount of his time teaching and throughout his writings there are references to teaching and to learning. He deals explicitly with teaching and learning in a number of his works, most notably in *De Veritate*, question 11 (Aquinas 1953); *Summa Theologica, prima pars*, question 117 (Aquinas 1948); and also *II Sentences* questions 9 and 28 (Aquinas 1929), though there are other passages where he discusses teaching and learning. It is clear that for Aquinas education has, and hence teaching and learning have, an unambiguous theological goal, namely, God, who is wisdom and truth. This theological dimension is articulated in the opening paragraphs of the *Summa Contra Gentiles*, where Aquinas says that the ultimate end of the whole universe is Truth and this is also the aim of the wise (Aquinas 1955).[5] This truth, he contends, is incarnated in the person of Jesus Christ.[6] Mindful that in the *Summa Contra Gentiles* he is not necessarily addressing Christian believers, Aquinas adds that Aristotle agrees that truth is the ultimate end of the wise (Aristotle 1976). That is, even if one does not begin from the position of someone who believes in God, in Aquinas' view, someone who seeks wisdom aims at the truth and there can be no further end than its attainment.

What emerges from the theological account that Aquinas gives is the recognition of the teacher as a role model for the pupil. The pupil learns by spending time with the teacher, not only listening to the words of the teacher, but by paying attention to his or her way of living out what he or she teaches.[7] It is thus important that the teacher be a person of good character, as the teacher inevitably serves as an exemplar for students. There is, therefore, an inescapable moral dimension to all teaching, and this is not restricted to the teaching of morals, but applies to other kinds of human knowledge. A teacher who loves his or her subject and who is enthusiastic is far more likely to capture and retain the attention of the learner than one who shows no commitment to the educative task.

Love and enthusiasm for the subject, while crucial to teaching, are not enough; the teacher must also genuinely care for the truth and be committed to possessing a mastery of his or her subject, so that he or she has the breadth and depth of knowledge requisite for confidently teaching his or her pupils. The pupil, on his or her part, begins by having faith that what the teacher is about to impart is trustworthy and that the teacher is knowledgeable about the subject. Faith is required not just for religious belief, but for scientific understanding as well, for as Aquinas says, we could not live in the world at all unless we are prepared to have faith.[8]

The emphasis on trust and faith in teaching and learning in particular highlights the importance of the relationship between the teacher and the learner. A poor or distant relationship will not facilitate learning, since it will not promote the trust required for the pupil to have confidence in the teacher. Aquinas emphasises the importance of friendship between teacher and pupil which develops a love of learning in the pupil. The learner must, if he or she is to grow in wisdom, listen willingly, seek diligently, respond prudently and meditate attentively.[9] In order for this to occur, the pupil needs to have the right conditions for learning, and a key component of these is the nurturing and encouragement that he or she receives from teachers.[10] This is in contrast to a ‘shopkeeper view’ of teaching and learning where there is no need for any relationship between teacher and learner, save for a commercial one in which a product is exchanged for financial gain. In such a view, learning is a transaction facilitated by the teaching of the teacher, a contractual obligation to be fulfilled. The educative process as Aquinas sees it is one which enables the relationship between teacher and learner to facilitate learning. Nevertheless, though trust is vital in the interaction between teacher and learner, he does not deny that there is something to be transmitted to the learner, but it is no inert product, the learner is actively involved in the learning process. There is an exchange between teacher and learner, but for Aquinas, this is a vitally active process, involving both teacher and learner. Teacher and learner are both engaged in a voyage of discovery for the truth. In this, he has much in common with the proponents of progressive education. Aquinas is fundamentally and critically interested in the question of how one person is able to teach another.

This is no facile question, but goes to the heart of the nature of teaching, since, despite the dividing of the concept of teaching into a task and achievement sense,[11] we still want to know whether what has been taught has been learned. It is achievement that matters to us. Hence, a teacher who attacks her subject with enthusiasm, but whose pupils fail to learn, is perhaps carrying out the task of teaching, but if the pupils fail to learn, she cannot be considered a successful teacher because nothing has been achieved. We are not satisfied simply by the adequate completion by the teacher of her teaching duties, that is, by her mastery of her subject, by her preparation of her lessons, and her performance of the act of teaching the pupils.

It is expected that learning will have taken place. A central question, then, in any evaluation of teaching is a concern for knowing whether the learner, as a result of teaching, now knows what the teacher knows. A classic puzzle, in the apparent absence of a mechanistic didactic process, is how teaching enables the learner to come to know what he or she did not know before, so that a new state in the learner is brought about. One response is to assert, as St. Matthew and St. Augustine do, that only God can teach, an assertion that Aquinas takes up the challenge to discuss and to explain.[12] Moreover, it might be added in support of this assertion, that it seems to be an uncontroversial empirical claim that one person cannot cause another to know, in the sense that it is a matter of efficient causation, where what the teacher does invariably leads to learning.[13] That is, experience tells us that sometimes despite the best efforts of a teacher, a student can fail to learn.[14] Teaching is often compared to an art[15] and Socrates, as is well known, compared teaching to the work of the mid-wife (Plato 1987, 25–29). Its success seems to be unpredictable and so it is possible to conclude that if learning occurs at all it is due to God's Grace. Aquinas agrees that in a sense only God teaches, but he also claims that teaching is an effective cause of learning.[16]

An immediate response we might make here is to question Aquinas about what he means by knowledge, that is, we need to ask what might be meant by saying that one person cannot cause another to know, or more positively, can cause another to know. It is plain that he does not think that learning is a mechanical process or that it takes place merely by means of signs, and it is evident that teaching is more efficacious than any other process in bringing about learning.[17] It is more efficacious, for instance, than allowing children to do what they like.[18] Whatever it is that is meant by knowledge and hence, coming to know what the teacher teaches, it is not a matter of a simple, straightforward transaction or a case of a simple operation of an efficient cause. What the student will have gained from the teacher will be an understanding of what is being taught, and this likely will not coincide with the understanding the teacher has of whatever is taught. For a start, the teacher has a wider understanding of the subject and sees connections that the pupil may not. Nevertheless, despite this apparent acknowledgement that to some extent, knowledge is constructed, Aquinas rejects a relativist view of knowledge and argues that human beings can discover the truth about the nature of the world and of themselves. That is, Aquinas rejects the view that individuals construct knowledge which is idiosyncratic, since the quest for knowledge is the quest for truth and whether something is true or not is not determined by individual whim.[19] He defends the notion of the individual human person as a being capable of intellectual knowledge, moral agency, and creative engagement with the world. Knowledge for Aquinas involves the use of our sensory and cognitive powers to gain an understanding of the interconnections among the phenomena that we experience.[20] In other words, the student always contributes something to what the student learns.

Although we can postulate a species of causal relation[21] between teaching and learning in that a pupil learns because the teacher teaches, it certainly can be concluded that the student in learning actualises something which did not previously exist, since it is not a straightforward replication of what the teacher knows. That is, there is a real change brought about in the world, for something new, knowledge in the student, has been created. What was not previously known by the pupil is now known, and the pupil is changed. Since only God has the power to create, to bring something into being *ex nihilo*, the issue of whether one person can teach another needs some explanation, since it seems apparent, that teachers can teach, that is, bring about learning, and pupils can know what they did not know before. In framing a response, Aquinas distinguishes between principal and instrumental causes, and argues that in a sense it is true that only God can teach, just as a pen can write provided there is an agent using it. By this, however, Aquinas does not intend to imply that human beings have no free will;[22] rather, his main point is that though human beings act as free agents in the world, they do so in co-operation with God, who is the source of all being. Nevertheless, if only God is able to bring something into existence *ex nihilo*, we need to be able to explain how that which was not known is now known, and how a real change has been brought about, since it is through teaching that something new now exists. The intellectual state of the learner is changed: if she has learnt, then she knows.

Aquinas argues that knowledge itself does not change, since knowledge is only knowledge if it is the truth, and what is true remains true, but there is a change from what is in potentiality to what is in actuality.[23] Underlying Aquinas' understanding of how teaching brings about learning is a conviction that the world is discovered, that realities previously unknown are brought to light. Finite beings do not create in the absolute sense, but they do co-operate in the unfolding of the universe and in that sense, they bring what was formerly only potentially known into actuality.

Austin Farrer provides a helpful account of the way in which human beings are co-operators in the fulfilment of God's plan for the universe and hence how they are able to act creatively. In one sense of 'cause,' it is true that God is the Cause of all things, so that human beings can only be a type of subordinate or secondary cause. Thus, though God is the ultimate author of all things, it does not mean that human beings are not able, in their own way, to act as causes. In acting as causes, human beings will not be directly aware of the Divine hand which is the source of their ability to act, but may be able through reason to apprehend its origin. Farrer comments that a person may suppose herself to infer God as the cause of the physical effects she studies, or as the cause of her own existence, without being aware of the divine causality behind her own thought. She may, in fact, be aware of it as a simply general illumination, lighting up all her understanding indifferently, so far as she understands; as a candle illuminates all equidistant objects with indifferent rays. In saying this, Farrer draws on Augustine's image of God shining through a person's acts of intelligence (Farrer 1948, 8–10).[24]

Farrer observes that God's actions are not apparent to human beings because we take for granted the light by which we see, the source of which may be hidden from us. Thus, it is possible to act without any consciousness of God acting through us. Farrer does not mean that we have no capacity to act through our own free will; rather, he means that it is through God's power that we are able to act at all. That is, when we act according to our nature and to the laws of nature, we are enabled to do so because it is through God's creative power that both our nature and the laws of nature exist, a point with which Aquinas agrees completely.[25] There is a distinction, says Farrer, between the First Cause and the secondary efficient causes. If God acts supernaturally, it is only for human beings that these acts are supernatural since for God, as the author of all being and whose nature is infinite, there are no actions which are not willed by Him and none which exceed His nature. For human beings acts may be designated as supernatural because they exceed human capacity to understand (Farrer 1948, 10).

Farrer explains that we should understand the term 'cause' as meaning an agent, and the term 'First Cause' as meaning a creative agent, which is not a cause in the first sense at all, since it is not an efficient cause. It is not, he says, to be thought of as a supreme causal law or as a first event from which other events follow. In a Kantian definition of cause, for example, a cause is an event belonging to a class of events, of which it is universally true that they are followed by events of a further given class. This is understood to mean that no cause is endowed with an efficacy beyond what it has in nature. What this implies, says Farrer, is that if we say that a flash of lightning is the cause of the consequent thunder, we are held to be classing the lightning as an electrical explosion, and acknowledging that from all electrical explosions sound-waves arise. If cause is understood in the Kantian sense, he says, then to talk of a cause being endowed with an efficacy beyond its natural scope is nonsense (Farrer 1948, 11–12).

That this is so is explained by Farrer in the following way. As already stated, according to this definition of 'cause,' no cause can be endowed with an efficacy above what it has by nature. This means that if event B follows an event A in a manner other than that which the causal law applicable to A demands, then by the Kantian definition A is not the cause of B at all, and B's cause must be sought elsewhere. If no natural cause for B can be established, then in view of the Kantian definition, it would not be possible to propose that what caused B was a supernatural event, since, by definition, this would mean the cause has an efficacy beyond what it has naturally. It seems that to avoid the unwanted violation of the definition we would have to say that the event was uncaused. The way is not open to us to attach the event to the First Cause, since it would not be then a natural cause. Furthermore, if the event is uncaused, then it seems to imply that it has simply come into being of its own accord, again, in violation of our definition of cause.[26]

What this shows is that difficult problems arise if it is supposed that God acts in the world in the same way that secondary causes, such as human beings, act. How God acts in the world remains unfathomable. There are hierarchies of human actions, where some are higher than others, as Aquinas says, but these, if they are the actions of human beings, remain within the natural world.[27] Nevertheless, Farrer points out that there is a two-sided aspect to our existence as human beings. On the one hand, we are active secondary causes in interaction with each other and, on the other hand, we owe our being to the first cause. We are never alone in the world, but always have our existence in these two ways, as active beings in the world and as beings with God. There is a sense in which both the human being and God are *en-act-ing* the human being's life, though in different ways and at different depths. The first cause operates in the secondary cause, says Farrer, and a number of difficult problems turn on this double agency.[28]

The idea of double agency allows us to see how, though God is the first cause of everything, He is not a cause in the usual sense, and so it is plausible to propose that human beings are secondary causes. Ultimately, God is the source of all creation, but double agency implies that human beings are able to collaborate in creation through what they can make. Their ability to act as secondary causes establishes their autonomy as distinct individuals, but still leaves us with the problem of how it is that knowledge is imparted to another. What has been shown, at least in outline, is the sense in which the assertion that it is God who teaches can be affirmed while not excluding our conviction that human beings teach. Moreover, it is also possible to see the sense in which we can say that human beings can create and how new knowledge can be discovered. It is the question of teaching itself to which we return in order to explain how one person is able to teach another.

Teaching, Aquinas argues, needs to employ, as far as possible, the same processes that the individual uses in coming to know anything at all about the world. The teacher leads the pupil to the knowledge of things unknown in the same way that one directs oneself through the process of discovering something one does not know.[29] In general, there are two processes that enable us to acquire knowledge and these are: (1) by discovery (*inventio*) and (2) by learning (*disciplina*). In the first case, we come to know through unaided natural reason, in the second, we are helped by a teacher. Aquinas says that in *discovery* the order of proceeding is this: first anyone who wishes to arrive at the knowledge of something unknown applies general self-evident principles to certain definite matters, second, from these moves to particular conclusions, and third, having done this, advances from these to others. Consequently, he says that one person is said to teach another if that individual is able to show the other person, through signs or general principles, the natural reasoning process that he or she used in arriving at those conclusions. Through having been led through that reasoning process, the pupil applying his or her own natural reason is able to come to know things that he or she previously did not know. Aquinas compares this process to that of a doctor who heals a patient, not through some power that she possesses herself but through the activity of the patient's nature. In the same way, Aquinas says, a teacher is able to cause another to have knowledge through the activity of the learner's own natural reason.[30]

Although Aquinas says that teaching takes place through the use of signs he explicitly rejects the idea that we learn through signs. Signs are instruments (*instrumenta*) which aid the learning process, but are no more than aids in the process of learning. In teaching another, the discourse of reason is expressed through signs (*per signa*) so that the student comes to know through these aids (or *instrumenta*).[31] It is through the principles which are represented by the signs that we learn. Aquinas says that to some extent we know the things we are taught through signs and to some extent we do not know them. Thus, he says, if we are taught what man is, we must know something about him beforehand, namely, the meaning of animal, or of substance, or at least of being itself, which last concept cannot escape us. Similarly, if we are taught a certain conclusion, we must know beforehand what the subject and the predicate are. Aquinas, in agreement with Aristotle (whom he quotes), suggests that learning comes from pre-existing knowledge.[32] For Aquinas, learning is an activity which starts from some pre-existing knowledge and proceeds through the use of reason to new knowledge.

Aquinas does not think that knowledge is innate, but quite sensibly says that we cannot teach someone who has no basic understanding of the general subject to begin with. Pupils need to be prepared to learn and to be in the right frame of mind before they are ready to learn. In respect to preparation, one kind of pre-existing knowledge, but not the only sort, that Aquinas has in mind are the general principles of logic. A second kind of pre-existing knowledge will be principles and concepts—signs represent both these—that are used to explain something new to a learner. Finally, learners also need to be in the right frame of mind, which means that they are paying attention and attending to the tasks of learning in an active way. Like Aristotle, Aquinas thinks that knowledge is *potentially* in the mind and has to be drawn into *actuality*, but it is also clear that the learner cannot be a passive vessel into which knowledge is poured.

Aquinas considers knowledge as being seeded, that there are *rationes seminales* (seminal reasons) which are immediately given and which arguably form the beginning principles from which knowledge can be built.[33] This seems to imply a kind of constructivism, since knowledge is built or acquired around these seeds, but this would be a mistaken view of Aquinas' position.[34] Aquinas says that we immediately know such things as the principle of non-contradiction, that the whole is greater than the parts, and that we should seek good and avoid evil.[35] This, however, does not commit Aquinas to constructivism and he explicitly rejects any relativist form of constructivism. He rejects it on the grounds that if the mind were to construct its own knowledge from sensory data this would imply that the mind already possessed that knowledge in actuality, since it would otherwise not be able to recognise the perception as being a perception of something. That is, in order to know that one sees a rose, one already has to know what a rose is.[36] Later, however, Aquinas nuances his position by saying that though it is true that the mind receives knowledge from sensible things, the soul forms in itself likenesses of things, inasmuch as through the light of the agent intellect the forms abstracted from sensible things are made actually intelligible so that they may be received in the possible intellect.[37] In saying this, Aquinas wants to steer a middle path between those who argue that knowledge is innate and only requires the senses to stimulate our minds into remembering, and those who argue for something like a naïve causal theory of perception, that is, that our knowledge is basically caused by external factors.[38]

There are common principles known immediately by the agent intellect in accordance with which our knowledge is constructed, but since these are available to everyone, it is possible to see how it is that human beings arrive at the same conclusions from the same sensory data. Every human being is equally human, but each human being individually possesses that human nature. In other words, though each human being has an individual rational nature, all human beings are recognisably similar in certain overlapping features. Although some writers argue that there is no common human nature, this does not accord with our experience,[39] since we are able to reliably recognise each other as human beings. Human beings do not each possess a human nature which is so radically different from that of another that we fail to recognise the other person as a human being. It is the possession of our distinctively human rational nature which enables human beings to come to know the truth, though each individual comes to the truth in his or her own way. Objective reality is something we subjectively learn and know in common with other human knowers. That is, knowledge for each person is obtained in the exercise of his or her own agent intellect and rational nature. Thus, Aquinas does not take a radical constructivist approach, but explains how knowledge is obtained in the mind in terms of act and potency.

The mind is related to external things in two ways. In the first way, things outside the mind are only potentially intelligible, that is, able to be known. The mind itself—at least that part of it which enables us to understand (viz., the agent intellect)—is active, since it is this power of the mind which makes potentially understandable things actually understandable. In other words, there is a part of the mind which acts to make sense of the information or sensory data we receive. In order for the mind to make sense of the sensory data it receives, it needs to situate these data where it can be acted upon. In the second way, the mind must be such that it can receive the sensory data which originate outside the mind. That is, sensory data are actualised by objects which are outside the mind, that is, scent of flowers comes from the flowers. Hence, external objects are the source of the sensory data that the mind receives. The part of the mind which acts as receiver Aquinas calls the possible intellect, that is, it receives the sensory data which are potentially knowledge and thus are made actual by the work of the agent intellect. That is, on receiving the sensory data which are, say, the scent of flowers, the agent intellect enables us to identify the sensory data as the scent of flowers and so we know that we are smelling flowers.

Aquinas notes that something can pre-exist in active completed potency, where something can bring about the thing into existence via an intrinsic principle. By 'pre-exist' Aquinas means that something is already present in a latent form, ready to act when the conditions are right for its action. The human immune system, for example, springs into action when it is needed; it is not activated until then. Hence, a person who is sick may, through the healing power of the body itself, be restored to health without any assistance from some external agent. Passive potency, on the other hand, means that something requires the aid of some external agent to bring it into actuality. For example, a doctor assists healing by administering medicines which act as instruments that restore health. Knowledge, by analogous reasoning, pre-exists in the learner, not as pure passive potency, but as active completed potency, that is, the seeds of knowledge already exist within us, which is to say, the capacity to learn and some basic understanding pre-exist in the active learner.[40] If this was not the case, a person could not acquire knowledge independently.[41] Just as the physician can aid the process of healing, so too can the teacher aid the process of learning. This can be done by utilising natural reason. Aquinas reiterates that knowledge gained exists in seminal form and can be developed by means of the activity of a created power.[42]

Instructively, he continues by remarking:

> We do not say that a teacher communicates knowledge to the pupil, as though the knowledge which is in the teacher is numerically the same as that which arises in the pupil. It is rather that the knowledge which arises in the pupil through teaching is similar to that which is in the teacher, and this was raised from potency into act, as has been said. (*DV* 11, Art.1, 85)

The interesting thing here is that Aquinas recognises that the knowledge gained by the pupil is not quite the same as that of the teacher, though of course, it cannot be entirely different, otherwise it could not be common knowledge or intersubjectively shared knowledge at all. This is the mistake made by the constructivist who claims that knowledge is constructed by the individual learner. If that were so, knowledge would be idiosyncratic and personal, and if we take knowledge to be public it could not be knowledge at all.

The constructivist is partially correct in that a learner does bring something to the learning process and the specific knowledge gained, namely the active potencies for learning. Moreover, and here is a further point where the constructivist is right, there is the proposition that the new knowledge that is gained has to become part of the learner's general understanding of the world. In other words, the learner has to make room in his or her general theory of the world for the new knowledge. If we think of this general theory of the world as part of an interconnected set of relationships between particular individual items of knowledge, then new knowledge needs to be inserted into this set of relationships and interconnections. As each individual has different experiences, then such new knowledge as is gained will be situated differently amongst the various items of knowledge that the person already has. Hence, the construction is of a new web of relationships amongst the items of knowledge that the person already possesses. The understanding of the interconnectedness of things will be different for each person. Some will see these webs of relations more deeply than others and one dimension of wisdom is born of the depth of understanding of these interconnections. The items of knowledge, because knowledge is about what is true, are the same for everyone; and in teaching, the first task is to enable pupils to learn what is true. The second and more difficult task is to convey to pupils how things interconnect. Pupils begin by learning facts of various kinds, such as the temperature at which water freezes and boils, the standard temperature and pressure under which this occurs, and so on. From facts about other kinds of liquids, a generalised theory about the interrelationships between temperature, volume, and pressure can be constructed. For pupils to understand this kind of interconnectedness between various quantities and qualities is to begin to learn about the world. The kind of interconnections between such physical qualities such as volume, temperature and pressure, is scientific knowledge, which Aquinas calls created wisdom. Much such scientific knowledge is, however, part of the legacy of those human beings who have blazed a trail before us and made discoveries about these interconnections. There is no need for us to perform the laborious task of making all of these interconnections ourselves. Since knowledge is communal, we can share in the community's accumulated wisdom. Knowledge of facts and their interrelationships leads to a third stage of learning wherein pupils fit what has been learned into their general understanding of the nature of the world. It is this third stage which depends on the individual, and could be said to be the individual's own construction. It is also where misunderstandings can occur and where the depth of our understanding can vary. It is also in the third stage where the possibility of wisdom which reaches below the surface of things can arise and albeit dimly, through becoming aware of relationships between objects, we may also come to discern the activity of the Divine Logos within the world.

3 Conclusion

Central to Aquinas' conception of teaching and learning is his recognition that the source of all knowledge and understanding is ultimately God. It is God who teaches in the primary sense, since it is through God we have our being. The senses play a crucial role in Aquinas' account of teaching, since it is through signs, not in themselves, but as understood in this instance as standing for underlying principles and for knowledge already gained, that teachers are able to convey to their pupils new knowledge and understanding. Before this is possible, however, learners need to be ready to learn and be prepared to actively engage in discovering new connections among things. They need to be prepared in two senses. First, they have to be in a state of readiness to learn and be at the right stage of psychological development. Second, they need to be readied by their teachers by means of, among other things, appropriate teaching settings, teaching materials, and learning cues. As learners—and as teachers—we are striving to know the truth, and this is a constant search for an ultimate understanding of how things connect together. It is not enough in the Thomist understanding of teaching and learning to have gained skills if these are not accompanied by some deepened understanding of how the skills acquired lead us closer to truth and so ultimately to God.

In his methodology, Aquinas is alive to both the transmission and facilitation models of teaching and learning. He proposes a middle way. He affirms the existence of a real world and the possibility of having knowledge of it. Moreover, in having knowledge, we know truth, and this has the practical consequence of enabling us to understand the world and to make the right kinds of decisions about our activities in the world. Since knowledge is about what is true and teaching can help us learn what is already known, there is a transmission sense in Aquinas' conception of teaching. At the same time, the centrality of experience in the learning process, leads him to also embrace the facilitation model of teaching in his conception of teaching. He advocates the use of the senses to discover what the world is like and teaching should as far as possible employ the same kinds of methods that the individual uses to discover things. Aquinas urges us to use all our capabilities to learn, for the end result of our learning should be the truth, and that Truth is God.

End Notes

1. For example, the Australian Catholic University (ACU) *Policy on Quality Teaching and Learning* (2006) states that its teaching and learning policy attends to "the spiritual, moral, values and ethical perspectives" (*sic*) and empowers staff and students to engage in teaching and learning that meets professional accreditation needs, is critical and well-informed, up-to-date with knowledge and research in the substantive disciplines, is innovative and makes appropriate use of information and communication technologies. It then lists 21 characteristics of effective teaching and 11 characteristics of learning promoted

by the "Learning Paradigm". What is not provided is a clear statement of the underlying philosophy of teaching and learning which itself is drawn from an articulation of a philosophy and theology of education. This is not to be critical of ACU, since few universities have a clear articulation of how they understand teaching and learning or even an awareness of the controversial nature of questions about teaching and learning. At URL: http://www.acu.edu.au/_data/assets/pdf_file/0003/98913/2009_Policy_on_Quality_Teaching_and_Learning.pdf, accessed 2 Sept 2009.

2. Peters quotes from the Faculty and TA Development Office at Ohio State University which proposes that a philosophy of teaching includes: (i) your conception of teaching and learning; (ii) a description of how you teach; (iii) justification of why you teach that way. See Peters (2009), 111–113.
3. Nola and Irzik describe the transmission model as the view that there is a fixed body of knowledge that has to be imparted to students. This model is widely criticised for assuming that there are objective propositions about the world and that these are what are to be taught and learnt. See Nola and Irzik (2005), 175. The facilitation model can be seen to have its roots in the Deweyan conception of learning by experience. See for example, Dewey (1938) and Kolb (1984). Aquinas proposes a middle way.
4. Sometimes so-called 'progressivists' use the term 'traditionalists' as a pejorative to describe teachers who employ teaching methods that stifle creativity and free expression in the classroom. For example, the curriculum that such traditionalists were said to have taught was highly structured according to the interests of the teacher (or school) and methods varied little, taking small account of the interests or the background knowledge and understanding that their pupils brought to the classroom. Progressivists, on the other hand, expressed an interest in the backgrounds of their pupils, seeking to engage them in learning by experience and as such are prepared to use a variety of methods, including allowing students freedom of expression. John Dewey, considered one of the founders of progressivism, was highly critical of some progressivists who saw progressivism as simply allowing pupils to do what they liked. For Dewey, freedom of expression as a pedagogical method did not mean lack of constraint. See Hirst (1974), 3–5 & 111–112. See also Dewey (1938), 1–11 and Dewey (1909), 24–25.
5. Aquinas 1955. *Summa Contra Gentiles* (hereafter *SCG*), 5 volumes, New York: Image Books, I, trans. A.C. Pegis, ch.1, #2. Aquinas says, "*Oportet igitur veritatem esse ultimum finem totius universi; et circa eius considerationem principaliter sapientiam insistere.*"
6. "*…ego in hoc natus sum, et ad hoc veni in mundum, ut testimonium perhibeam veritati.*" ("…for this I was born, and came into the world, that I might bear witness to the truth.") (*Jn* 18:37).
7. On this point, Boland says that more attention should be paid to what Aquinas says in his Gospel commentaries. Boland (2006b), 299. See also Boland (2006a)
8. Aquinas says, "*…quia si homo nollet credere nisi ea quae cognosceret, certe non posset vivere in hoc mundo. Quomodo enim aliquis vivere posset nisi*

crederet alicui? Quomodo etiam crederet quod talis esset pater suus? Et ideo est necesse quod homo credat alicui de iis quae perfecte non potest scire per se." ("…if a person was only willing to believe that which he knew himself, he would certainly not be able to live in this world. How can someone live without believing anyone? How would he even believe that this man was his father? It is necessary that a person believes someone about what he cannot by himself know perfectly.") See Aquinas (2006), *Proemium*, 3.

9. Aquinas, *Sermon Puer Iesus*, at URL:http://www.op-stjoseph.org/Students/study/thomas/SermPuerIesus.htm. Accessed: 31/7/09. See also Boland (2007b)
10. The letter, *De Modo Studendi*, attributed to Aquinas, but held to be of dubious authenticity, captures some of the right conditions. For example, some of the practical suggestions made are: try to reach difficult things by means of small steps, ready your mind through prayer, try to be friendly to everyone, and listen to good teachers. See Aquinas (1951). See also Torrell, who suggests that the text was not written by Aquinas: Torrell (2005, 360).
11. Ryle originally introduced the task/achievement distinction in reflecting that some verbs have a task sense, for example, "He ran the race" and "he won the race". In Ryle's sense the former is a task sense and the latter is the achievement sense. Teaching in its task sense can be understood as that which is required to be carried out in order for something to be considered under the concept of teaching. Hence, planning of lessons, delivering the lesson and so on can be understood as the task sense of teaching. The achievement sense of teaching is the satisfactory performance of the tasks. Another sense of the task of teaching, which we are considering here, is the idea that the task of teaching involves the instruction of learners in order that they learn. The achievement sense of teaching involves the idea not only that the tasks listed have been carried out, but that the learners have learned. See Ryle (1949) and Peters (1966), 36–27.
12. *Matt.* 23: 8, which is quoted in Aquinas' introduction to Question 11 in *De Veritate* (hereafter *DV*) 11, Art. 1, 77. This passage is also discussed by Augustine in *De Magistro* (*The Teacher*). It is perhaps the central lesson of the work. See Augustine (1955), 94 n. 1.
13. Aquinas argues against the view that human beings share a common passive intellect as several awkward conclusions follow from assuming a common passive intellect. For example, since the immortal part of each human being is common, there does not seem to be any reason for anyone to strive to be virtuous, nor would it be strictly true to say: 'This man knows'. Moreover, if humans did share a common passive intellect, then teaching would only need to activate what is already there. This would suggest that there could be a simple causal process that could unlock the knowledge that is common to all human beings. Instead, Aquinas defends the notion that each person has his or her own individual passive intellect. See Aquinas (1968).
14. Aquinas is well aware of this. The student needs to be receptive to learning and there is a developmental order according to which human beings mature. Following Aristotle, Aquinas recommends an order in which areas of knowledge should be

taught to students. These are outlined in his *Commentary on Causes*: "first logic which teaches the method of the sciences, then mathematics which even young people can learn, next natural philosophy which requires time in which people can gain experience, only then moral philosophy which is a subject to which a younger person cannot be properly receptive (*cuius iuvenis esse conveniens auditor non potest*) (of which a young person cannot be a suitable student), and finally divine science which considers the first causes of things." Aquinas (1996), Preface.

15. See for example Highet (1989).
16. By effective cause is meant an efficient cause. Aquinas says that the teacher causes knowledge in the learner by reducing him from potentiality to act. *ST* I.117.1.
17. *DV* II, Q. 11, Art. 1, #2, 84. In asserting this, Aquinas draws on the discussion in Augustine's *De Magistro* of whether words are signs. Signs do play a part in aiding us to learn, but not just by themselves. See Augustine (1995), 97–103.
18. Some progressivists might argue against this, but we shall not pursue the point here.
19. Here we should distinguish between constructivism as learning theory which emphasises the importance of individual differences in teaching children and constructivism which argues that each individual uniquely constructs his own knowledge. That is, in the latter case, knowledge generation is determined to a large extent by social factors and so stands in opposition to the idea of a mind-independent world that human beings can access through observation and the use of reason.
20. Stump's account of Aquinas' epistemology suggests that it is a form of externalism with reliabilist elements. Such an interpretation is supported by the account of teaching and learning in Aquinas presented here. See Stump (2003), 235. For an account of externalism and reliabilism in epistemology see Armstrong (1973).
21. Although we have said that it is possible that no learning takes place even though there was teaching, this does not entail that there is no causal relation between teaching and learning, just as it is possible that a kettle fails to boil at 100 °C. In both cases, we would look for other causes. Teaching is not a sufficient cause of learning.
22. Aquinas argues that humans do have free will. Otherwise, exhortations, punishments and rewards would have no point, and moreover, it is clear that humans are able to exercise some control over their desires and appetites. *ST*, I, 83, Art. 1; see also Aquinas (1962). *Peri Hermeneias: Aristotle on Interpretation Commentary by Thomas Aquinas finished by Cardinal Cajetan*, trans. J.T. Oesterle, Milwaukee: Marquette University Press (also known as *De Interpretatione*, hereafter *DI*), Book I, Lesson 14, para. 18.
23. Aquinas is a realist about knowledge. Through the intellect, we come to know things as they are. Knowledge, he says, pre-exists in the learner potentially in the sense of an active potency. We shall return to this below. *DV* 10, 4, 19–20 and *DV*, 11, 1, 83.

24. See Augustine (1976). *De Civitate Dei* (*The City of God*), Harmondsworth: Penguin, Book X, Ch. 2.
25. Aquinas says, "the divine will must be understood as existing outside of the order of beings, as a cause producing the whole of being and all its differences." *DI* I, Lesson 14, para. 22.
26. It is a poor definition of cause that rules out existential dependency relations by definitional fiat because the Kantian definition is undermined.
27. *ST* I-II.1. Arts. 1, 5 and 6. See also McInerny (1993), 196–216.
28. Farrer is mainly thinking about theological problems here, but there are also philosophical problems, including the nature of free will. Farrer (1948), 28–33.
29. *DV* 11, art.1, 83.
30. *DV* 11, art.1, 83.
31. Boland remarks that this seems to indicate the influence of Augustine. Boland (2007a), 48.
32. Aquinas (1970), Book I, Lect. 1 (*Posterior Analytics*, I, 1, 71a,1).
33. *DV* 11, Art. 1, #5, 84–85.
34. Boland seems to suggest that Aquinas leans towards constructivism. See Boland (2007a), 46–47. See also n. 19 above for a brief definition of constructivism.
35. *DV* 10, Art.12, #3, 67.
36. He says (of those who would make an inferior cause the complete source of our knowledge):

> Other proponents…said that the soul is the cause of its own knowledge. For it does not receive knowledge from sensible things as if likenesses of things somehow reached the soul because of the activity of sensible things, but the soul itself, in the presence of sensible things, constructs in itself the likenesses of sensible things. But this statement does not seem altogether reasonable. For no agent acts except insofar as it is in act. Thus, if the soul formed the likenesses of all things in itself, it would be necessary for the soul to have those likenesses of things actually within itself. This would return to the previous opinion [that knowledge is innate] which held that the knowledge of all things is naturally present in the human soul. (*DV* 10, Art. 6, 27–28)

As mentioned earlier, Aquinas rejects the Averroist view that human beings have a common agent intellect as well as a common possible intellect. This entails that individuals gain knowledge and understanding through their own individual agent intellect and possible intellect. (Aquinas, *De Unitate Intellectus Contra Averroistas*) Given that knowledge is public, we need a means of accounting for agreement amongst people about what it is that they are talking about. It was plain enough to Aquinas that people were able to communicate with one another. If the mind were to construct its own knowledge from sensory data it could only do so by using what it already knew, otherwise, we would have no basis for claiming that what we had constructed was knowledge. The mind does not construct knowledge in this way, though it is the case that knowledge is gained through the activity of the agent intellect.

37. "And in this way all knowledge is in a certain sense implanted in us from the beginning (since we have the light of the agent intellect) through the medium of universal conceptions which are immediately known by the light of the agent intellect. These serve as universal principles through which we judge about other things and in which we foreknow these others. In this respect, that opinion is true which holds that we previously had in our knowledge those things which we learn." *DV* 10, Art.6, 28.
38. This is perhaps most clearly seen when Aquinas says that the mind has contact with singulars, that is, particular sense data, through the mediation of particular reason, a power of the sensitive part, that is, via the brain. Hence, knowledge is not caused directly by objects stimulating our senses. *DV* 10, Art.5, 23.
39. See for example: Rorty (1998), 167–185.
40. *DV* 11, Art.1, 82. Just as when a physician administers medicine to speed up the natural healing process. This is in fact a mixed or more complex situation than appears—a person may have sufficient power of his own to bring about healing, but needs assistance, so it is not entirely passive potency here. That is, the natural healing processes are capable on their own to restore health, but medicine in this case, speeds the process. In other cases, only medical assistance can restore health.
41. Aquinas considers and rejects the idea that someone can be his or her own teacher, but this does not mean that he or she cannot acquire knowledge for himself or herself. Aquinas agrees that one can learn things by discovery, but he says that only a teacher will have understanding of the entire subject or science and so can teach it more easily. *DV* 11, Art. 2, 89–91.
42. *DV* 11, Art. 1. 83–85.

References

Aquinas, Thomas. *Sermon Puer Iesus*, at URL: http://www.op-stjoseph.org/Students/study/thomas/SermPuerIesus.htm. Accessed 31 July 2009.

Aquinas, Thomas. 1929. *Scriptum super libros Sententiarum* (*In I–IV Sentences*). Tomes I-II ed. P. Mandonnet, Lethieulleux, Paris, Tome III (1933) and Tome IV (1947) ed. M.F. Moos.

Aquinas, Thomas. 1948. *Summa Theologica*. Trans. Fathers of the English Dominican Province. New York: Benzinger Bros., republished 1981, Westmoreland: Christian Classics.

Aquinas, Thomas. 1951. *De Modo Studendi*. Trans. V. White. Blackfriars, Oxford: Oxonian Press.

Aquinas, Thomas. 1953. *Truth* (*De Veritate*), vols. I–III. Trans. R.W. Mulligan, J.V. McGlynn, and R.W. Schmidt. Chicago: Henry Regnery Company.

Aquinas, Thomas. 1955. *Summa contra gentiles*, 5 vols. New York: Image Books.

Aquinas, Thomas. 1962. *Peri Hermeneias: Aristotle on interpretation commentary by Thomas Aquinas finished by Cardinal Cajetan*. Trans. J.T. Oesterle. Milwaukee: Marquette University Press.

Aquinas, Thomas. 1968. *De Unitate Intellectus Contra Averroistas* (*On the unity of the intellect against the Averroists*). Trans. B.H. Zedler. Milwaulkee: Marquette University Press.

Aquinas, Thomas. 1970. *Commentary on Aristotle's posteriorum analyticorum*. Trans. F.R. Larcher. Albany: Magi Books.

Aquinas, Thomas. 1993. *Commentary on Aristotle's Nichomachean ethics*. Trans. C.I. Litzinger, Foreword by R. McInerny. Notre Dame: Dumb Ox Books.

Aquinas, Thomas. 1996. *Super Librum de Causis Expositio* (*Commentary on the book of causes*). Trans. V. Guagliardo, C.R. Hess, and R.C. Taylor, Intro. by V. Guagliardo. Washington, DC: Catholic University of America Press.
Aquinas, Thomas. 2006. *Collationes: In symbolum apostolorum, De decem praeceptis et Expositio salutationis angelicae*. Pamplona: Editio Catholica.
Aristotle. 1976. *The ethics of Aristotle*: *The Nichomachean ethics*. Trans. J.A.K. Thomson and Hugh Tredennick, Intro. by J. Barnes. Harmondsworth: Penguin.
Armstrong, D. 1973. *Belief, truth and knowledge*. Cambridge: Cambridge University Press.
Augustine. 1976. *De Civitate Dei (The city of God)*. Harmondsworth: Penguin.
Augustine. 1995. *Against the academicians, the teacher*. Trans., Intro. and notes by P. King. Indianapolis: Hackett.
Boland, V. 2006a. What happens when minds meet? Thomas Aquinas on the mystery of teaching and learning. *Doctrine and Life* 56(6): 3–17.
Boland, V. 2006b. Truth, knowledge and communication: Thomas Aquinas on the mystery of teaching. *Studies in Christian Ethics* 19(3): 287–304.
Boland, V. 2007a. *St. Thomas Aquinas*. London: Continuum International Publishing Group.
Boland, V. 2007b. St. Thomas's sermon *Puer Iesus*: A neglected source for his understanding of teaching and learning. *New Blackfriars* 88(1016): 457–470.
Dewey, J. 1909. *Moral principles in education*. Boston: Houghton-Mifflin.
Dewey, J. 1938. *Experience and education*. New York: The MacMillan Company.
Farrer, A. 1948. *The glass of vision*. London: Dacre Press.
Highet, G. 1989. *The art of teaching*. New York: Vintage Books.
Hirst, P. 1974. *Knowledge and the curriculum*. London: Routledge & Kegan Paul.
Kolb, D.A. 1984. *Experiential learning*. Englewood Cliffs: Prentice-Hall.
McInerny, R. 1993. Ethics. In *Cambridge companion to Aquinas*, ed. N. Kretzmann and E. Stump, 196–216. Cambridge: Cambridge University Press.
Nola, R., and G. Irzik. 2005. *Philosophy, science, education and culture*. Dordrecht: Springer.
Peters, R.S. 1966. *Ethics and education*. London: George Allen & Unwin.
Peters, M. 2009. Editorial: A teaching philosophy or philosophy of teaching? *Educational Philosophy and Theory* 41(2): 111–113.
Plato. 1987. *Theaetetus*. Trans. R. Waterfield. Harmondsworth: Penguin.
Plato. 2005. *Protagoras and Meno*. Trans. A. Beresford, Intro. by L. Brown. London: Penguin.
Rorty, Richard. 1998. *Truth and progress*. Cambridge: Cambridge University Press.
Ryle, G. 1949. *The concept of mind*. London: Hutchison.
Stump, E. 2003. *Aquinas*. London: Routledge.
Torrell, J-P. 2005. *Saint Thomas Aquinas*: *The person and his life*, vol. 1. Trans. R. Royal. Washington, DC: Catholic University of America Press.

Aquinas on Connaturality and Education

T. Brian Mooney and Mark Nowacki

Meno: But how will you look for something when you don't in the least know what it is? How on earth are you going to set up something you don't know as the object of your search? To put it another way, even if you come right up against it, how will you know that what you have found is the thing you didn't know?

Socrates: I know what you mean. Do you realise that what you are bringing up is the trick argument that a man cannot discover either what he knows or what he does not know? He would not seek what he knows, for since he knows it there is no need of the inquiry; nor what he does not know, for in that case he does not even know what he is to look for.

Meno: Well, do you think it is a good argument?

Socrates: No. (*Meno*, 80d3-e5)

1 Introduction

St. Thomas Aquinas, the Angelic Doctor of the Catholic tradition, has had a massive influence on the philosophical and theological teachings of western Christianity. Nevertheless, at no stage did he formulate an explicit philosophy of education. He infrequently uses the term *educatio* (education), and when he does, he is usually appealing to the ancient and fundamentally Platonic conception of *paideia* (educational formation).[1] Aquinas uses *educatio* most often in connection with the notion of parental responsibilities in regard to the raising of children.[2] Aquinas thus echoes the Platonic position that education should be thought of within a very broad context, the basis of which is the formation of a rational and socially responsible soul.

T.B. Mooney (✉) • M. Nowacki
Singapore Management University, Singapore, Singapore
e-mail: brianmooney@smu.edu.sg; nowacki@smu.edu.sg

T.B. Mooney and M. Nowacki (eds.), *Aquinas, Education and the East*, Sophia Studies in Cross-cultural Philosophy of Traditions and Cultures 4,
DOI 10.1007/978-94-007-5261-0_3, © Springer Science+Business Media Dordrecht 2013

As Plato expresses it: "By *paideia* I mean virtue (*arete*) in the form in which it is acquired by a child." (*Laws*, 653b)

Aquinas' theory of education is formulated within a general metaphysical account of human nature (what is nowadays called a philosophical anthropology) and, in particular, an account of human rationality that is grounded in human nature which in turn provides us with the possibility conditions for our moral language rising to the level of making claims that are both meaningful and objectively true. We give a brief synopsis of the metaphysical narrative Aquinas provides, but in a contemporary key that draws upon the resources of Analytical Thomism.[3]

We then turn to an overlooked epistemological issue that arises as a consequence of the ontology we defend. Our discussion is inspired by the traditional Thomistic understanding of *connatural knowledge*. We provide a description of our situated position as human knowers that displays how recognizing connatural knowledge is not only necessary for a robust epistemology but also for progress in accounting for how we become educated as moral knowers and agents. In brief, we address the problem of moral education by considering how, given that we are the sort of beings we are, we need to cultivate certain habits. These habits are required for human flourishing and for successful moral perception, so that we become able to recognize salient features of moral situations for what they are.

Our exposition of the Thomistic ontology of the human person, together with the notion of connatural knowledge, yield a defensible account of human flourishing that can undergird meaningful and true claims in morality. The central idea is that moral perceptiveness is directly relevant to understanding of moral scenarios and thus our capacity to characterize them accurately. If we lack the necessary perceptiveness we will lack a full understanding of the relevant moral concepts and hence will inevitably fall short in understanding and doing what we ought. In moral education we need the capacity to see, to have an appropriately schooled affectivity, and to have the sort of character that is properly responsive to moral salience, in order to act as we ought.

By proceeding in this way, we hope to give due weight to Aquinas'—and Aristotle's[4]—contention that moral philosophy is in itself a rather useless thing. As Aquinas writes in the *Disputed Questions on the Virtues in General*:

> It should be said that…prudence implies more than practical science. Practical science [i.e., moral philosophy] makes a universal judgment of things to be done, for example, fornication is evil, theft ought not be committed, and the like. This knowledge can be present yet reason's judgment concerning the particular act be intercepted with the result that one does not judge correctly. That is why moral science is said to avail little for the acquisition of virtue, because even when it is had a man can sin against virtue. (Q. 6 *ad* 1) (Aquinas 1999, 37–38)

Understanding concepts—especially moral concepts—involves more than appreciating their dictionary definition. Often our concepts have no meaning for us unless we not only have a grasp of the dictionary definition but also have the sorts of inclinations (forms of affectivity) we should have upon presentation of objects falling under that concept. Generosity is only truly understood by the generous person; Scrooge didn't get it until he had it. The buying of the Christmas goose consolidated Scrooge's nascent understanding of generosity into a generosity understood and put into act as a habit of character.

It is well understood in the Christian tradition that at times we may well understand the nature of a moral concept but fail to be moved to act in a moral way. This need not be simple incontinence or weakness of will, what the Greeks called *akrasia*.[5] For instance, we may fail to do what we ought because we are tired or suffering from some general ennui. At other times we may be better placed to act because we are energized. In fine, recognition creates one set of difficulties; failing to act on situations that are properly recognized is another. The fact that we share a common moral language is not enough. We can be taught to engage in philosophical discourse about moral concepts, but an ability to manipulate abstract concepts is sufficient neither for recognition nor understanding. Failing to recognize morally salient features of a situation, and to be properly affected by those features, involves much more than a lack of adequate vocabulary or even basic competence in drawing formal connections among moral concepts.[6]

It is in this broader context, in which understanding requires recognition and recognition ultimately requires experience in action, that we should talk not so much about connatural knowing as connatural doing or connatural acting. Or so we shall argue presently.

2 Philosophical Anthropology and Its Ontological Background in Aquinas

As we have pointed out, Aquinas' philosophical anthropology is itself dependent upon broader metaphysical theses. Human beings and other natural kinds are the sorts of things they are because they are substances possessed of natures that manifest the natural necessities that apply for them. There are several crucial moves that explicate this perspective. Aquinas' metaphysics is first of all teleological and in its turn this teleology is grounded in what contemporary parlance would call natural necessity. So, we begin with a contemporary version of the Thomistic story. The central assumptions we make in what follows are these: first, we live in a universe of *things*. Our universe is populated by entities that persist over time, and we will call these things 'substances'. Second, we can directly observe how substances change and interact over time, often in predictably consistent ways. From these parsimonious resources we now construct a contemporary substance-metaphysics.

Natural necessity is a type of necessity found *in rerum natura* that flows from the intrinsic natures of substances.[7] *What* a thing is determines both how it exercises active power and how it can be passively acted upon by other things. The operations of a substance progress in accordance with its nature; and it is through the operations of a substance that we discover, *a posteriori*, what its nature is. "In short, the relation between what a thing is and what it is capable of doing or undergoing is naturally necessary" (Harré and Madden 1975, 14).

For example, a woman bringing her finger into contact with a rose thorn pricks herself and sheds a drop of blood. The sharpness and hardness of the thorn are consequent upon the nature of the rose. The penetrability of the woman's flesh follows from her having the nature of a human being, since the nature of human beings is such that a woman has soft outer flesh (thus differing from, for example, an oyster). If one encountered a thorn that could not prick a finger, or a finger that could not be pricked by a thorn, then one should doubt that it was a *rose* thorn or a *human* finger.

The illustrations of natural necessity common experience furnishes are legion. Ice, but not copper, melts at 0 °C. Unlike cats, bluebirds fly.[8] As Sarah Waterlow observes, the behavior of a substance "is never a function of external conditions alone. All change…is at least partially determined by the subject itself, and in this sense there must be inner principles of change if there is change at all" (Waterlow 1988, 27). It follows that since the operations of a substance are determined by its nature, a substance cannot retain the same nature while it loses its usual capacities and powers. For illustration, if a piece of gold ceases to be malleable, we should have compelling grounds for thinking that, since it has different capacities and powers, the material has undergone a change in nature. Gold *qua* gold is malleable, whereas granite *qua* granite is not. Should a Humean object that it is logically possible for a continuously existing sample of gold to suddenly be transformed into granite, we reply that this thin logical possibility is no counterexample. It is not the properties of *the gold* that have changed, only that granite has been *substituted* for the gold. Such a substitution is logically, not naturally, possible. Given *what* gold is, gold does not naturally have the capacity to turn into granite and still remain gold. This is because gold and granite are different kinds of substances. It is *a priori* evident that one cannot modify features that are necessary to a thing's being the kind of thing it is if it is to remain the same kind of thing. But how do we know that gold and granite count as two different natural kinds? Harré and Madden answer succinctly:

> We follow the scientific tradition in identifying the real essence of a kind, material or individual, with its nature, which is progressively revealed *a posteriori* by empirical investigation. The real essence is only finally discovered when the analysis is complete. So far as we can tell, no analyses of … substances have yet been completed, so our knowledge of the natures of things is as yet an approximation to the knowledge of their real essences, but can stand in for that knowledge in all relevant contexts. (Harré and Madden 1975, 102)

All substances that share a range of causal powers constitute a natural kind. The accumulation of empirical data has prompted the redrawing of certain natural kind boundaries. Until the advent of modern chemistry two distinct kinds of substance were lumped together under the name *jade*: true jade, or jadeite, which is relatively rare; and nephrite, which is comparatively common. The distinction between true jade and nephrite is apparent only upon chemical analysis. Expansion of our knowledge of causal powers—in this case, a range of naturally necessary chemical reactions—revealed the diverse natures of these two substances.[9]

It is naturally necessary that we possess the human nature that we have. Our nature leads us to manifest forms of activity conducive to human flourishing. This grounds the fact that teleological thinking is not only embedded in our everyday discourse but resonates with our empirical experience of the world. It is impossible to adequately describe an action without appealing to some end or goal. The end may be external to the action, as the city of Singapore is external to us as we journey towards it, or internal to the action, as when we learn for the sake of learning.

For our present purposes we focus on living things, beginning with *parts* of organisms. We know that a hand is for grasping and that teeth are for chewing. It is not just for convenience that we identify distinct organs in living things. But to be an organ is to have a function, as the root meaning of the word reveals. A stomach is for digestion, and a heart is for pumping blood; these are *inter alia* their purposes. Such examples show the presence of purpose that is recognized, not merely imposed, by our understanding.

We also discover purposes at the level of the *total* organism in five ways—some depending upon others. First, we can see that these purposes must be present because without mentioning them we cannot accurately describe the subsidiary purposes of the parts. That *teeth are for biting* is something we can recognize only by understanding that some animals partially fulfill their holistic good by their ability to bite. We know the purposes of parts, but this presupposes our understanding—albeit incomplete—of how those parts function within a broader way of life.

Second, we can be certain that we know the holistic good of organisms because we know what is bad for them. For instance, we know that it is bad for a dolphin to be bitten by a shark. But this involves knowing what is good for the dolphin—at the very least the preservation of bodily integrity, which partly constitutes a good life for it.[10]

Third, having a sense of the holistic good of organisms embedded in a way of life allows us to recognize what sorts of things must be present for the purposes of whole organisms to be realized more or less fully. This yields the notion of a flourishing life. A well-fed dolphin is a more flourishing dolphin than a less-well-fed one. But recognizing a more flourishing dolphin involves recognizing much that is distinctive about a dolphin's way of life, notably that they carry out activities such as hunting, feeding, and raising their young in an essentially social way.

Fourth, knowing what counts as a flourishing life for an organism allows us to take a broadly normative perspective on what that organism needs for a flourishing life, in other words what a thing *ought* to have if it is to flourish. This is a hypothetical necessity. *If* dolphins are to flourish *qua* dolphins, *then* they must have water to swim in. The conditions necessary for flourishing can also embrace forms of behaviour constitutive of ways of life. This is clear not only in the case of dolphins, but for all social animals, from ants to bees to human beings. Flourishing for a social organism may include its place within a hierarchy and the cognitive abilities it needs for communication as well as abstraction from experience.

Fifth, knowing what counts as a flourishing life for an organism allows us to identify features, including psychological ones, that enhance or undermine flourishing. Although animal nature may be plastic and responsive, there are more and less successful adaptations. Learning to swim or use simple tools brings more food within a primate's reach, and so enhances its flourishing. On the other hand, a loss of confidence can cripple a young chimpanzee's chances of acquiring a mate.[11] Just as dolphins need water and sustenance, so too do humans need air and food. And, as we will now argue, for humans flourishing includes the acquisition of those reflectively appropriated and refined habits we call virtues.

3 Virtues and Connatural Knowledge

Given the metaphysical story just presented, we can begin to spell out which teleological orderings are naturally necessary for human beings to flourish. In brief, it is naturally necessary for human flourishing that we develop habits conducive to such flourishing. The relevant habits here are, of course, virtues, and we will treat only of those virtues that are widely accepted as being connected with flourishing for human beings.[12] After discussing virtues, we will probe more deeply into their formation, and in particular we will look into the educationally significant notion of connatural knowledge (that is, affective knowledge or knowledge by inclination).

Through experience and education human beings are socialized into the virtues. A *virtue*, as Aristotle and Aquinas teach us, is a stable disposition (i.e., a habit, or more properly, a *habitus*),[13] ideally deliberated, based on a mean relative to ourselves, and determined by reference to the practical reason displayed in the judgments and actions of the *euphronimos* (i.e., the person possessing excellence in practical wisdom).[14] For example, a virtue such as courage is a habit in the form of a stable disposition to behave in a certain way in the face of danger based on a mean between a state of deficiency (such as timidity) and a state of excess (such as foolhardiness) relative to ourselves (since people display courage in different ways and degrees) and determined by reference to the way a paradigmatically courageous person would judge and act. In a slightly looser but more contemporary idiom, virtues are embodied heuristics, rules of thumb guiding effective practical action that have been coded into human subjects via a process of training and (ideally) reflective appropriation.

Virtues run deep. As Aristotle suggestively puts it, virtues form a second nature within us.[15] Virtues are tools of human flourishing,[16] making their possessor good and rendering good his or her work.[17] They allow their possessors to make appropriate and effective choices within the practical sphere even in the absence of formal deliberation. Virtues do so by inclining the virtuous individual both to perform actions of a certain type and to develop the sort of character that will be a font of actions of this desired type.[18] A just individual is inclined to just acts, and a courageous individual is inclined to courageous acts. The thematic nature of actions that virtuous individuals undertake can thus be seen as supportive of particular ways of life.[19]

It is important to understand that virtues are not situated only at the level of the individual but are developed and manifested within a social context. For instance, the practical coherence of a set of virtues possessed by an individual is constantly tested against the socially-embedded individual's understanding of the world. Reinforcement comes in the form of both positive and negative social feedback. As John Dewey argues:

> Honesty, chastity, malice, peevishness, courage, triviality, industry, irresponsibility are not private possessions of a person. They are working adaptations of personal capacities with environing forces. All virtues and vices are habits which incorporate objective forces. They are interactions of elements contributed by the make-up of an individual with elements

> supplied by the out-door world. They can be studied as objectively as physiological functions, and they can be modified by change of either personal or social elements.... But since habits involve the support of environing conditions, a society or some specific group of fellowmen, is always accessory before and after the fact. Some activity proceeds from a man; then it sets up reactions in the surroundings. Others approve, disapprove, protest, encourage, share and resist. Even letting a man alone is a definite response. Envy, admiration and imitation are complicities. Neutrality is non-existent. Conduct is always shared; this is the difference between it and a physiological process. It is not an ethical "ought" that conduct *should* be social. It *is* social, whether bad or good. (Dewey 2002, 16–17)

There is, then, an inherently social dimension to the virtues. Insofar as particular kinds of acts are valued by a group, the group will habituate its members in those virtues that lead to performance of valorized acts. And what sorts of acts will be valorized by a group? Clearly, those sorts of acts to which the virtues possessed by group members incline them. Our social group schools us in the virtues we ultimately develop and in the light of whose possession we either flourish or whither as human beings. Thus, it matters much not just what we are taught, but who teaches us, when they teach us, and why they teach us what they do. We will come back to this social dimension of habituation later.

We have seen that virtues are naturally necessary for human beings including *qua* social beings. Following Aquinas we now show how virtues involve *connaturality* in general and *connatural knowledge* in particular. For ease of exposition we begin with a discussion of connatural knowledge, and return to the broader notion of connaturality later. Connatural knowledge is knowledge that is readily acquired by beings of a certain nature.[20] For instance, dogs have knowledge of a scent-world that exceeds our capacity, because dogs are by nature better suited to process olfaction. There is a strong connection between connatural knowledge, which is a kind of knowledge by inclination, and the acquisition and successful possession of the virtues. This is because, as a kind of knowledge by inclination, connatural knowledge can be used to explain both how we recognize the morally salient features of situations as well as how we have inclinations to specific types of action in those recognized situations. Ultimately, it is only because human beings possess the specific nature they do that they are able to form certain moral concepts.

Aquinas variously calls connatural knowledge "judgment by inclination" (*per modum inclinationis*, *S.T.* I.1.6); "affective cognition" (*cognitio affectiva*, *ST* I.64 and 97, II-II.162) and "experiential cognition" (*cognitio experimentalis*, *ST* II-II.97). It is a species of cognition which involves both apprehension and judgment. Sometimes Aquinas talks of connatural knowledge as a sort of judgment: "He who has the habit of charity judges rightly of such matters by a sort of connaturality." (*ST* II-II.45) At other times he talks of it as *receptiveness* to correct judgment, inclining us towards objects of desire and love: "Because where there is the greater charity there is the more desire; and desire in a certain degree makes the one desiring apt and prepared to receive the object desired. Hence he who possesses the more charity, will see God the more perfectly, and will be more beatified." (*ST* I.12)

Whichever reading one adopts, Aquinas thinks that our possession of virtues *inclines* us towards salient features of what we connaturally know, so that we may *recognize* and *judge* what is to be done both practically and intellectually. Connatural *knowledge* involves both apprehension and judgment. A dog has connatural apprehension of noises that we cannot hear. This metaphysical fact is reflected in epistemology. Since the ability to reliably discriminate an X (a certain noise) from non-Xs entails possessing the concept of an X, then given John Searle's observation that no subject can hold beliefs that embody concepts which that subject fails to have (Searle 1992, 155–162), it follows that unlike us, the dog may form the judgment that the noise is present.

Connatural knowledge itself is dependent upon the broader notion of connaturality. Connaturality refers to those specific metaphysical accidents readily acquired by beings because of their nature—initially, their first nature, which is the sort of substantial being they are. In what follows we introduce an original distinction between two types of connaturality. In the first place there is the connaturality that belongs to animals (human and non-human) and other living things *qua* beings of a certain sort. We might term this *ontological connaturality* since it is our first nature that provides the setting within which the receptiveness and aptness of the knower may be actualized. The natural inclinations of human beings towards goods such as life itself and those conditions required to sustain it, extend to goods such as intellectual curiosity and sociability. These inclinations are ontologically connatural since they provide our reason with its premises and with criteria for making sense of experience.[21]

What we call *habitual connaturality* involves our second natures, that is, our first natures as suffused with virtues. Aquinas appeals to this notion when he writes:

> rectitude of judgment is twofold: first, on account of perfect use of reason, secondly, on account of a certain connaturality with the matter about which one has to judge. Thus, about matters of chastity, a man after inquiring with his reason forms a right judgment, if he has learnt the science of morals, while he who has the habit of chastity judges of such matters by a kind of connaturality. (*ST* II-II.45.2)

Habitual connaturality is acquired through the practice of virtue and involves a directed perceptiveness awakened by the possession of the virtue in question. The knowledge arising from this form of connaturality need not be propositional. Rather, it may be a form of *knowing how* to perceive the salient moral features in circumstances that demand the exercise of a virtue. For example, a friend may know how to perceive one's sadness which would be overlooked by a non-friend. The apprehension of sadness calls for moral judgment about what needs to be done.

To bring out the complexity and practical depth of these issues, note that both types of connaturality that we distinguish entail a role for love. Moreover, since love engages us at the core of our ethical being, both sorts of connaturality involve forms of affectivity that shape us as moral agents.

Ontological connaturality entails love because it involves an aptness that is both an *apprehending* and a *relating*. For, as Aquinas points out, "Love is said to discern because it moves the reason to discern" (*ST* II-II.47.1 ad 1). Like all loves, the lover is related to the beloved, but full comprehension—as opposed to apprehension[22]—of the beloved may well be missing. In ontological connaturality the objects of our connatural affectivity incline us towards them and make us receptive to their attractions.[23] There is an identification of sorts, though not yet a fully reflective union, obtaining here between lover and beloved.

Habitual connaturality likewise entails a role for love. It involves an understanding that is embedded in the acquisition and practice of specific virtues. This understanding occurs because the subject identifies with the object of cognition, and this identification is love-in-act. A virtuous person not only does the right thing but takes delight in—or *loves*—doing the right thing because it is the right thing to do.[24]

The role of love and presence of affectivity more generally highlight a key feature of connatural knowledge. Whether our connatural knowing is grounded immediately in ontological connaturality, or whether it is immediately grounded in our habitual connaturality and only mediately in our ontological connaturality, in both cases there is always the presence of inclination. However, inclination is not necessarily present in propositional knowledge, even when propositional knowledge refers to matters significant to practical reason. A smoker who accepts the truth of the proposition that smoking is bad may still continue to smoke. But a person who accepts the truth of the proposition that she should act appropriately on her generosity would not be generous if she regularly fails to do so.

To see the interplay of identification, love, affect, and inclination, consider the following two examples. First, contrast pressing a button that causes hundreds of people to die in a city 1,000 miles away but without ever seeing the results of the act with running someone through with a sword. Second, contrast undergoing the physically intrusive surgical procedure of having an abortion with taking the morning after pill.

In both examples there are significant differences in the ways one understands (as opposed to merely knows) the moral dimensions of one's actions. Pressing a button is very different from running a sword through a person before you. When running someone through it is impossible not to experience at first hand the pain and horror of another's death together with the appropriate feelings and sense of the moral gravity of the act. Similarly the physical procedure of clinical abortion has immediate responses at the level of affect that taking a pill does not seem to generate. Moral sensibility is not just appropriating accurate propositional descriptions of the situation. Rather, it involves the whole person, including her inclinations and affectivity. It is one thing to have propositional knowledge of a moral act. It is quite another to understand that act by means of identification at the depths of one's fully engaged being. Thus the inclinations and affects that accompany certain actions are partially constitutive of our understanding of morally significant actions.[25] Without this interplay among identification, love, affect, and inclination, the moral salience and appropriate apprehension of moral gravity are vitiated.[26]

Having explored the ontological and habitual dimensions of connaturality, we are now in a position to draw out some further thoughts on how issues of identification, inclination, love, and affectivity inform our agency as mediated by our first and second natures. It is naturally necessary that we develop habits (our second nature) because human beings, who are by nature agents, necessarily express their first nature in activity. Through appropriate action comes the development of a habit for that sort of action. As Aquinas sums up Aristotle, "like actions produce like habits."[27] But our second nature must be in line with our first nature for us to flourish. For example, as a matter of ontological connaturality, we are naturally inclined towards sociability, and we may know this by comparing ourselves with other non-sociable animals such as great white sharks. Unlike us, a female great white shark might be content to devour her offspring. Such is not typically the case for humans. As a matter of ontological connaturality, we are naturally inclined towards parenting. But humans may be better or worse at parenting, with ensuing degrees of flourishing. Good parenting, put in constant practice, becomes a habit and ultimately a skill. *Qua* habit, skilled parenting is part of our second nature and includes virtues of attentiveness, care, and concern. These virtues are partially constitutive of excellence in human parenting.

The formation of virtue engenders a correlated capacity for perception of practical and moral salience. A good parent is capable of perceiving potential dangers that may be overlooked by a non-parent. For instance, a parent would spot staples on the floor that a baby could consume. A non-parent would likely overlook the threat or would not be immediately concerned to pick up the staples. It makes little sense to say that one is disposed to act in one way rather than another unless one is able to recognize the salient features of situations that trigger one's acting upon that disposition.[28]

Aquinas addresses this issue in his *Commentary on the Divine Names*:

> experiencing [*passio*] seems to pertain more to the appetite than to cognition, because things known are in the one knowing after the manner of the person who knows them and not after the manner of the things which are known, but the appetite moves one to the things as they are in themselves and thus he receives an affection for the things themselves. Just as a virtuous man, by the habit of virtue which he has in his appetite, is perfected to judge rightly about what pertains to that virtue, so he who has an affection [*afficitur*] for divine things, receives divinely the right judgment about divine things.[29]

Aquinas teaches us that we need to be formed in virtuous habits that dispose us to recognize moral saliency and provide us with an impetus for action, ideally as an immediate consequence of recognition. For in most circumstances, it is best that there be no gap between *seeing* and *doing*. Consider G.E.M. Anscombe's telling discussion on "affected ignorance" (Anscombe 2005, 65). As she notes, for most of us the homeless do not exist because they are, in a real sense, invisible to us. Only those who have been habituated to perceive the homeless will see them as they are and, after perceiving them, be moved to alleviate their plight. The virtuous agent sees and does, and those who would not do, do not see.

Understanding at the level of our whole engaged being requires the formation of virtues. The virtues are partially constitutive of what it is to be rational. We are most rational when our affections, our inclinations, our loving volitions, and our cognitive identifications are all in play; to exclude any of them results in poor choices and less reliable practical action. Reason is important because we have (a) affections, inclinations, loving volitions, and cognitive identifications; and we can have (b) good affections, good inclinations, good loving volitions, and good cognitive identifications; but (c) all of these may or may not be under our rational control.

We are often but not always inclined towards *prima facie* goods for us *qua* human beings—not always, as the habits formed in our second nature may be vicious. Moreover, we might not act, for example, on our habitual inclinations due to trying circumstances, inattention, and other practical impediments. More generally, we are ontologically connaturally attracted to the good of pleasure. Pleasure is a good thing, for it moves us towards and reinforces our pursuit of ends that are good for us. But not every pleasure is unqualifiedly good. So our natural inclinations need to be supplemented by our reason to determine whether, and to what degree, a specific pleasure contributes to our flourishing. This is because while there is nothing wrong with pleasure itself, we may take delight or forge pleasurable associations with actualities that are not in themselves suitable for us to pursue given our first nature as human beings. When this happens, the habits informing our second nature are vices. In contrast, the habitual connatural knowledge found in moral and intellectual virtues reveals, by directed inclinations, those interests we have *qua* human beings and lead us to our distinctively human sort of flourishing. These remarks apply *mutatis mutandi* to our affections, loving volitions, and cognitive identifications.

4 Applications to Contemporary Education

There is a startling unanimity among philosophers who acknowledge the role of connatural knowing (either explicitly or implicitly) when it comes to how they view education in general and moral education in particular. Plato, Aristotle, and Aquinas, as well as the philosophical traditions they represent, hold that:

> one who is intent on virtue should have some sort of moral training from his earliest years that he may rejoice and be sorrowful about the right things. This is proper instruction for youths so that they become accustomed to take pleasure in good works and be grieved in evil works. Therefore, teachers of youth compliment those who do good and reprove those who do evil.[30]

If Aquinas' account of the centrality of proper habituation and connatural knowledge is correct, then much of what goes on in our contemporary classrooms is highly suspect. We cannot call ourselves responsible teachers if we over-specialize the teaching of virtue to the point where it becomes a class that students take once a week. What sort of moral formation are students undergoing, as they are obliged to sit quietly at desks throughout the school day, only to sit quietly at tables studying for the rest of the day, rarely being exposed to the sorts of practical challenges and experiences that would afford them the scope they need to grow into moral adults?

It is appropriate now to reflect on what implications the positions just developed have for contemporary education. In what follows, we will focus upon education in Singapore, but our remarks concerning the Singaporean educational system can, in many cases, be generalized to other countries as well. There are two reasons for this. First, Singapore is acknowledged globally as a centre of excellence in education. Second, Singapore, like many other countries, has inherited a British model of education. A salient feature of the British educational system is that it is test based. Performance on examinations whose grading schemes involve application of internationally standardized rubrics determine a student's academic future.[31] Understandably, tests focus upon the consolidation of considerable factual knowledge and standard methodologies and techniques.

In Singapore, the educational system focuses almost exclusively on the impartation of know-that—knowledge of facts—instead of know-how—skills that emerge as excellence in practice.[32] Connatural knowledge, which is clearly based upon experiential learning, is a form of know-how raised to the level of a skill—as indeed are all the virtues in the fullest sense. Also missing in the Singaporean educational context is a robust cultivation of know-why.[33] Again, virtues are apt to involve know-why: as in the case of the generous individual who wishes to perform a generous act precisely because it is the generous thing to do.[34]

Let us first contextualize our discussion by noting how Singapore has developed institutional structures to support its educational goals. First, and most obviously, the entire educational infrastructure is geared towards producing pupils who excel on standardized tests. Students are constantly tested, constantly examined throughout their scholastic careers. Admittedly, all this practice in test-taking means that Singaporean students generally have excellent test-taking skills. They know how to do well in exams.

The prevailing pedagogical approaches found within Singapore schools are naturally consistent with the realities of this exam culture. In the classroom a *transference model* predominates. The teacher lectures, students take copious notes, and the teacher routinely tests to check whether the contents of the notes are recalled accurately. More drill and reinforcement is usually necessary, so the majority of students in Singapore take some form of private tuition, and spend long hours outside of school going over what they were supposed to have learnt within it.[35] The controlling metaphor here is of pouring water from a full jug into an empty one: the teacher is active and giving, the student passive and receptive.

But no teacher, however gifted, could ever experience for the student what the student must himself or herself experience in order to learn. Connatural knowing requires *experiences* of the relevant sort to sensitize learners to the content and salient features of its targeted subject matter. Without the requisite experiential grounding, a student at best achieves a formal conceptual or linguistic facility. And this is precisely what the Singaporean model of education is in danger of delivering.

For instance, in an elite primary school, one of the authors has helped introduce experiential lessons to support the school's pastoral care programme. Such a lesson would typically involve a brief introduction and discussion of an ethical concept, which is then followed by an activity (usually a game). The lesson concludes with discussion and reflection on how the targeted ethical concept was embodied in the experience. Student contributions during discussion followed a distressing pattern. First, the students (who were 11 or 12 years old) would show great intellectual facility. They could manipulate concepts and draw relevant inferences from them. So, from the claim 'Everyone should act fairly' students could infer 'She ought to act fairly.' However, during the activity, students would routinely manifest behaviour at odds with their earlier claims. For instance, students found it difficult to operationalize the concept of turn-taking—they routinely failed to see the importance of not obstructing or interfering with the turns of other players—and even broke game pieces when they disagreed with other players performing legal moves that were to their disadvantage. When later queried as to whether their own actions qualified as 'fair', students always claimed they were. On one memorable occasion, when asked, 'So what does *fair* mean?' a female student replied, in perfect seriousness, 'Fair means other people giving me what I want.' No wonder this child is happy to say that other people should be fair! It is clear that what these children lacked was not intelligence (students in the pastoral care programme could expect to qualify for places in academically elite secondary institutions) but rather the sorts of reflective experiences necessary for formation in the virtues. To use Alasdair MacIntyre's language, these students not only failed to perceive goods internal to practices (i.e., playing fairly) but they lacked a capacity to properly recognize the practice itself (i.e., fair playing) (MacIntyre 1985). The classroom thus needs to be a place in which students are provided with experiences that enable them to understand and operationalize virtuous activity so as to cultivate genuine know-how and connatural knowing.

The variety of children's life experiences in Singapore is minimal.[36] To call a Singapore youth a 'student' and to call that same youth a 'child' amounts to much the same thing. To be young is to be in school, and to be in school is to spend the majority of one's day in a highly structured environment that, as noted above, rewards passivity.[37]

With truncated life experience comes truncated opportunity to develop basic social virtues. Virtues that would normally develop in youth may develop, but often with a different emphasis than expected or in an incomplete form. Courage is replaced by bare endurance. In Singapore the master virtue is, with good reason, tolerance, but tolerance conflicts with objective considerations of value and does not necessarily require the sort of affective participation and understanding required to make objectively formed judgments. Hence, despite the prevalence of so-called Confucian values in Singapore one encounters an extraordinary level of moral relativism masked as tolerance. In its turn this means that the sort of holistic educational practice that Aquinas advocates is missing. A flourishing human rationality requires the proper engagement of our affections, our inclinations, our loving volitions, and our cognitive identifications. In short, a varied diet of experiences is urgently required for the realization of the full potential of students.

In the absence of a richer experiential pedagogy, we are likely to succumb to a kind of schizophrenia wherein the social value of tolerance is elevated beyond the point at which it can be reasonably sustained. Social cohesion in Singapore may come to be valued more than truth. Moreover the methods and institutional standards of success assumed in the education process may lead not so much to questioning but to tolerant silence—a far cry from the Thomistic disputation. Philosophies of education must be carried out in the *agon*—the field of conflict. Critical evaluation, searching, and questioning, and not bare appeal to authority, are required.[38] Such a pedagogy embraces the whole person and is responsive to the demands for flourishing made at both our first and second natures. We need to be schooled at the levels of affect, inclination, loving volition, and cognitive identification, and in doing so we must always appeal to the rich tapestry of experience in which connatural knowing and its concomitant virtues find their home.

End Notes

1. On St. Thomas' use of *educatio* see for example his *Disputed Questions on Truth*, Q. 11 (*De Magistro*); *Summa Contra Gentiles* (hereafter *SCG*) III.122; *Commentary on the Sentences* IV. For translations and sources see respectively Aquinas 1953, 1975, and 1852–1873.
2. Indeed, Aquinas uses the terms *nutritio* (nurture) and *education* interchangeably in *SCG* III.122 (Aquinas 1975).
3. The term 'Analytical Thomism' was originally coined by John Haldane, who describes it as involving "the bringing into mutual relationship of the styles and preoccupations of recent English-speaking philosophy and the ideas and concerns shared by St Thomas and his followers" (Haldane 2004, xii). For a useful overview see Paterson and Pugh (2006).
4. See Aristotle, *Nicomachean Ethics* II.4, 1105b (Aristotle 2002).
5. For Aristotle's classic discussion see *Nicomachean Ethics* VII (Aristotle 2002).
6. Moral education has to amount to more than learning a language, even when we expand the notion of learning a language to include immersion in ways of life. As Alasdair MacIntyre has taught us, this is because ways of life themselves need critique, a critique which might not be available from a purely internal perspective. The Polynesian notion of the taboo which MacIntyre cites is a classic example. See MacIntyre (1985, 110ff).
7. We owe much in what follows to Harré and Madden, who discuss natural necessity in detail. See Harré and Madden (1975).
8. Even in the counterfactual world of comic books Superman has to be acknowledged as a Kryptonian because, however much he may look like a human being, his powers fail to track the same set of natural necessities found in humans.
9. This also works in the other direction. It took chemical analysis to reveal that rubies and sapphires are both members of the mineral species corundum.
10. We owe this example to Alasdair MacIntyre (2001, ch. 3).
11. It is also possible to train an animal to behave in ways that undercut its own flourishing, as the experiments by Seligman demonstrate concerning learned helplessness for dogs. See Seligman (1975).
12. The remarks we make concerning virtues apply, *mutatis mutandis*, to vices, for vices also are habits of moral weight.
13. The English word 'habit' can be misleading because the focal sense of 'habit' is something that we have reduced to a mechanical routine and which we possess best when we have to think about it the least. The Latin *habitus*, however, retains the notion of a formed disposition (like the English 'habit') but additionally a *habitus* is (ideally) something with which we are intellectually engaged. In this paper we will use the English 'habit' to refer to *habitus*. For further discussion

see John of St. Thomas, *Cursus philosophicus*, IV Q. 18 'On the Categories', especially art. 3 (John of St. Thomas 1883); for the distinction between 'habit' and *habitus*, see Simon (1986, 48–61).

14. See *Nicomachean Ethics* II.6 (Aristotle 2002) and Aquinas' commentary on the same passage (Aquinas 1993).
15. See *Nicomachean Ethics* II.1 and VII.10 (Aristotle 2002), *Rhetoric* I.11, *Magna Moralia* II.6.
16. As the pragmatist philosopher John Dewey puts it, "virtues are ends because they are such valuable means" (Dewey 2002, 47).
17. Thomas Aquinas, *Commentary on Aristotle's Nicomachean Ethics*, Bk. 2, lect. 6 (Aquinas 1993).
18. See Aquinas, *Commentary on Aristotle's Nicomachean Ethics* Bk. 2, lect. 2, par. 264 (Aquinas 1993).
19. A nice, common-sense example is given by Dewey:

> In practical life, there are many recognitions of the part played by social factors in generating personal traits. One of them is our habit of making social classifications. We attribute distinctive characteristics to rich and poor, slum-dweller and captain of industry, rustic and suburbanite, officials, politicians, professors, to members of races, sets and parties. These judgments are usually too coarse to be of much use. But they show our practical awareness that personal traits are functions of social situations (Dewey 2002, 19–20).

20. See *Summa Theologica* I.13.1 *ad* 8 (hereafter in text as *ST*) (Aquinas 1981); (Anscombe 2005, 59–66; Maritain 1953; Simon 1986, 48–61).
21. Aquinas and the natural law tradition generally maintain that our first nature furnishes us with the primary precepts of the natural law. These precepts and their ontological foundations are not discovered or justified by rational argument in which we move from premises to conclusions. Instead, they provide the very basis for our reasoning (especially practical reasoning) and are *per se nota* (i.e., immediately and directly known). Connatural knowledge is thus grounded in ontological connaturality. See Aquinas, *Commentary on the Sentences* III. d. 33, q. 2, a. 4, qc. 4; d. 37, q. 1, a. 1, ad 3; d. 37, q. 1, a. 3; *ibid.* IV, d. 33, q. 1, a. 1 (Aquinas 1852–1873); *De veritate* q. 10, a. 6, ad 6; q. 11, a. 1; q. 11, a. 3; q. 16, a. 1 (Aquinas 1953); *Summa contra Gentiles* III, 46 (Aquinas 1975); *ST* I.79.12; *ST* I-II.90.4 ad 1; 93.2 s.c.; *ST* I-II.100, a. 3; *ST* II-II 47.6 (Aquinas 1981).
22. We can apprehend a book from its cover but we do not comprehend it until we have read it carefully.
23. Or, contrariwise, the objects of our ontological connatural affectivity may repulse us as being unsuitable and we find ourselves naturally unreceptive to them.
24. As an aside, it is traditionally supposed that the unity of the virtues in Aristotle and Aquinas is achieved through the master virtue of practical wisdom (*phronesis* or *prudentia*). If our account of connaturality is correct it may well be that the deeper ground of the unity of the virtues should be sought in an account of love.
25. For related discussion of these points see Tallon (1997).

26. Interestingly, this deeper engagement of our being involves the person at the level of physical embodiment. For example, speaking in a neuroscientific context, Antonio Damasio writes that when a:

> bad outcome connected with a given response option comes into mind, however fleetingly, you experience an unpleasant gut feeling....What does [this] *somatic marker* achieve? It forces attention on the negative outcome to which a given action may lead, and functions as an automated alarm signal which says: Beware of danger ahead if you choose the option which leads to this outcome. The signal may lead you to reject, *immediately*, the negative course of action and thus make you choose among other alternatives.... Somatic markers probably increase the accuracy and efficiency of the decision process. Their absence reduces them....In short, *somatic markers are a special instance of feelings generated from secondary emotions*. [Secondary emotions are emotions that arise at the level of our second nature, like shame.] Those emotions and feelings *have been connected, by learning, to predicted future outcomes of certain scenarios*. When a negative somatic marker is juxtaposed to a particular future outcome the combination functions as an alarm bell. When a positive somatic marker is juxtaposed instead, it becomes a beacon of incentive (Damasio 1994, 173–174).

So both inclination and affectivity are important to our full appreciation of moral concerns. The affects that accompany certain actions—including somatic effects characteristic of certain emotions—are part and parcel of our understanding of the moral significance of actions. It is easier to do wrong when our feelings are not engaged.

27. *Commentary on Aristotle's Nicomachean Ethics* Bk. 2, lect. 1, par. 253 (Aquinas 1993). See Aristotle, *Nicomachean Ethics* II.1, 1103a14-b1 (Aristotle 2002) and Aquinas' commentary on the same (Aquinas 1993).
28. As McInerny points out, the recent resurgence of interest in virtue ethics "has been prompted by the realization that, apart from an acquired appetitive disposition, moral knowledge cannot be effective. Our moral character is a condition of our ability to see the demands of the good in particular circumstances. 'As a man is, so does the end appear to him.' [*Qualis unusquisque est, talis finis ei videtur*]." (Ralph MacInerny, in his Preface to Aquinas 1999, xviii.)
29. *De Divinis Nominibus*, chap. 2, 1.4, 191–192. Translated in Naus (1959, 147). Naus' comments are particularly relevant here: "But the prime analogate of connatural knowledge appears to be that of affective connaturality, in which the appetite affects the judgment of reason. Joseph de Finance, S.J., has analyzed the metaphysics which lies at the basis of affective connaturality in a context of determining moral values. The suitability of an object is recognized much more often through concrete relations of the appetite, by way of inclination, than through a purely rational and cold knowledge" (Naus 1959, 142–143).
30. *Commentary on Aristotle's Nicomachean Ethics* Bk. 2, lect. 3, par. 268 (Aquinas 1993).
31. The Singapore educational system institutionalizes the practice of high-stakes testing in an extreme form. In addition to the internationally recognized GCSE O-level and A-level examinations, students in Singapore are required to pass a national Primary School Leaving Examination (PSLE). Performance on the PSLE

determines, first, whether a student qualifies for secondary school; second, which secondary school the student may enter; and third, which academic stream the student qualifies for within a secondary school. For most Singaporean students, whether they will ultimately attend university is settled at the age of 12. Much useful information is available on the Ministry of Education's official website: http://www.moe.gov.sg, accessed 20 May 2011. For more on the PSLE, see http://en.wikipedia.org/wiki/Primary_School_Leaving_Examination, accessed 18 May 2011. Of course, high-stakes testing is hardly unique to Singapore; for further discussion see Lingard (2010, 129–147); and Au (2008). The impact on pedagogical practice, as teachers find themselves 'teaching to the test,' have also been studied: see Cuban (2008) and McNeil (2000). See also Chua (2009).

32. This reality is widely recognized and has become a hotly contested political issue. For instance, the manifesto of the Singapore Democratic Party, on its page dedicated to education, prominently displays the following observation by Roger Schank: "You don't have a great education. Your sense of a well-educated man is someone who has memorized all the facts." (http://www.singaporedemocrat.org/manifestoeducation.html, accessed 2 May 2011) (Schank 2011).
33. Know-why is concerned with uncovering causes, ends, and goals; with identifying that for the sake of which something is done, undertaken or pursued, or holds true. There has been official acknowledgement that the educational culture of Singapore does not support the sort of open-ended inquiry necessary for students to appropriate the meaning or meanings of what they study. As the former Minister for Education, Tharman Shanmugaratnam expresses it, what the system needs is "to create a culture of reasoning and discussion that nourishes children's natural sense of intellectual curiosity" (Shanmugaratnam 2006, 16).
34. For a full discussion of the distinctions and interpenetrations of know-that, know-how, and know-why, see Mooney and Nowacki (2011, ch. 1).
35. It is common for upper primary level students in Singapore to spend 10 h per day in school and for secondary school students to spend 12 h. The vast majority of Singapore students also take private tuition after school and on weekends; this often amounts to an additional 8 h of study per week without taking into consideration time spent on homework.
36. Anecdotally, it may be worth pointing out that university educators often note that Singaporean tertiary students are significantly less mature and less vocal than their international counterparts.
37. On a positive note, it may be remarked that introducing experiential learning methods into the Singaporean educational system can generate significant positive returns. For example, in a recent study it was shown that introducing a structured game-based thinking skills programme led to significant improvements in student performance across all English-medium subjects. See Pelizzo (2010).
38. The academic benefits of cultivating a philosophically informed approach to learning are well-supported within the educational literature. For a survey of approaches that have been empirically tested and have shown good results, see Millet and Tapper (2011).

References

Anscombe, G.E.M. 2005. In *Human life, action and ethics: Essays by G.E.M. Anscombe*, ed. Mary Geach and Luke Gormally. Charlottesville: Imprint Academic.

Aquinas, Thomas. 1852–1873. *Scriptum super quatuor libros sententiarum Magistri Petri Lombardi*. Parma: Pietro Fiaccadori.

Aquinas, Thomas. 1953. *Truth*, vol. II. Trans. James V. McGlynn. Chicago: Henry Regnery Co.

Aquinas, Thomas. 1975. *Summa Contra Gentiles*. Trans. Anton C. Pegis et al. Notre Dame: University of Notre Dame Press.

Aquinas, Thomas. 1981. *Summa Theologica, complete English edition in five volumes*. Trans. Fathers of the English Dominican Province. Allen: Christian Classics.

Aquinas, Thomas. 1993. *Commentary on Aristotle's Nichomachean Ethics*. Trans. C.I. Litzinger, foreword by R. McInerny. Notre Dame: Dumb Ox Books.

Aquinas, Thomas. 1999. *Disputed questions on virtue: Quaestio Disputata de Virtutibus in Communi and Quaestio Disputata de Virtutibus Cardinalibus*. Trans. Ralph McInerny. South Bend: St. Augustine's Press.

Aristotle. 2002. *Nicomachean Ethics*. Trans. and intro. by Christopher Rowe, intro. and commentary by Sarah Broadie. Oxford: Oxford University Press.

Au, Wayne. 2008. *Unequal by design: High-stakes testing and the standardization of inequality*. New York: Routledge.

Chua, Jude Soo Meng. 2009. Saving the teacher's soul: Exorcising the terrors of performativity. *London Review of Education* 7(2): 159–167.

Cuban, Larry. 2008. *Hugging the middle: How teachers teach in an era of testing and accountability*. New York: Teachers' College Press.

Damasio, Antonio. 1994. *Descartes' error: Emotion, reason, and the human brain*. New York: Grosset/Putnam.

Dewey, John. 2002. *Human nature and conduct*. Mineola: Dover Publications.

Haldane, John. 2004. *Faithful reason: Essays Catholic and philosophical*. London/New York: Routledge.

Harré, Rom, and E.H. Madden. 1975. *Causal powers: A theory of natural necessity*. Oxford: Basil Blackwell.

John of St. Thomas (Ioannes a S. Thoma). 1883. *Cursus philosophicus Thomisticus*. 9 vols. Paris: Ludovicus Vivès.

Lingard, Bob. 2010. Policy borrowing, policy learning: Testing times in Australian schooling. *Critical Studies in Education* 51(2): 129–147.

MacIntyre, Alasdair. 1985. *After virtue*, 2nd ed. London: Duckworth.

MacIntyre, Alasdair. 2001. *Dependent rational animals*. La Salle: Open Court.

Maritain, Jacques. 1953. *The range of reason*. New York: Charles Scribner's Sons.

McNeil, Linda. 2000. *Contradictions of school reform: Educational costs of standardized testing*. New York: Routledge.

Millet, Stephan, and Alan Tapper. 2011. Benefits of collaborative philosophical inquiry in schools. *Educational Philosophy and Theory*. doi:10.1111/j.1469-5812.2010.00727.x.

Mooney, T.Brian, and Mark Nowacki. 2011. *Understanding teaching and learning*. Exeter: Imprint Academic/St Andrews Studies in Philosophy and Public Affairs.

Naus, John E. S.J. 1959. *The nature of the practical intellect according to Saint Thomas Aquinas*. Rome: Libreria Editrice Dell' Università Gregoriana.

Paterson, Craig, and Matthew S. Pugh (eds.). 2006. *Analytical Thomism: Traditions in dialogue*. Aldershot/Burlington: Ashgate.

Pelizzo, Riccardo. 2010. *Report on explicit teaching of analytical thinking skills through games-based facilitation for all courses (in primary and secondary schools) for higher academic achievement*. Singapore: Ministry of Education, Singapore.

Schank, Roger. 2011. Singapore democratic party, *Manifesto: Education*. http://www.singaporedemocrat.org/manifestoeducation.html. Accessed 2 May 2011.

Searle, John. 1992. *The rediscovery of the mind*. Cambridge: MIT Press.
Seligman, Martin. 1975. *Helplessness: On depression, development, and death*. San Francisco: W.H. Freeman.
Shanmugaratnam, Tharman. 2006. Opening address. In *Philosophy in schools: Developing a community of inquiry, report on the proceedings of the conference*, ed. Ho Wah Kam and Henry Widjaja. Singapore: Singapore Teachers' Union.
Simon, Yves. 1986. In *The definition of moral virtue*, ed. Vukan Kuic. New York: Fordham University Press.
Tallon, Andrew. 1997. *Head and heart: Affection, cognition, volition as triune consciousness*. New York: Fordham University Press.
Waterlow, Sarah. 1988. *Nature, change, and agency in Aristotle's physics: A philosophical study*. New York: Oxford University Press.

Aquinas and the Second Person in the Formation of Virtues

Andrew Pinsent

1 The Mystery of Aquinas' Virtue Ethics

The study of the virtues has enjoyed a remarkable revival in contemporary ethics, a development that is, in part, a response to the sense that ethics cannot be reduced to the determination of right actions with respect to a set of rules or likely consequences. It is a commonplace of daily life that we judge persons, and not only actions, as being 'just', 'unjust', 'generous', 'mean', 'honest', 'dishonest', and so on. Furthermore, most people value ways to form those good traits called 'virtues' and diminish those bad traits called 'vices', at least in others and certainly in the young.

As MacIntyre has observed, the contemporary revival of virtue ethics has taken its main inspiration from the Greek classical tradition, with Aristotle as its principal focus and the *Nicomachean Ethics* as its 'canonical text' (MacIntyre 2007, 257, 259, cf. 147). Given the prominence of Aristotle to this revival, one would expect that Aquinas also has a great deal to offer contemporary scholarship on the virtues. Besides his commentary on Aristotle, *Sententia libri Ethicorum* (In NE), Aquinas produced an extraordinary volume of writing on his own account. The *Summa theologiae* (*ST*), Aquinas' largest and most influential work, covers his famous five proofs for the existence of God in just a single article (*ST* I.2.3), an article that has attracted enormous scholarly interest. By contrast, there are 1,004 articles, approximately one third of the *ST*, devoted to systematic accounts of the virtues and associated matters.[1] Yet even this material does not exhaust all that Aquinas has to say about the virtues, given the existence of other, smaller works on the virtues together with pertinent material across a wide range of other philosophical and scriptural commentaries.[2] The volume of text alone suggests that the study of Aquinas on the virtues ought to be a fruitful endeavor for philosophers today.

A. Pinsent (✉)
Faculty of Theology, University of Oxford, Oxford, UK
e-mail: andrew.pinsent@hmc.ox.ac.uk

T.B. Mooney and M. Nowacki (eds.), *Aquinas, Education and the East*, Sophia Studies in Cross-cultural Philosophy of Traditions and Cultures 4,
DOI 10.1007/978-94-007-5261-0_4, © Springer Science+Business Media Dordrecht 2013

Yet Aquinas' virtue ethics has presented a puzzle to contemporary philosophers and his contribution, to use MacIntyre's words, has been "unexpectedly marginal" (MacIntyre 2007, 178). The essential problem is one of interpretation. Despite Aquinas' great respect for Aristotle, there are many ways in which Aquinas' account of the virtues is not Aristotelian, despite manifest influences on many specific issues.[3] For example, one would expect any philosopher in the Aristotelian tradition to regard philosophical wisdom (*sophia*) as the summit of the virtues. Indeed, the prominence accorded to the Aristotelian *sophos*, sometimes translated as 'sage', has parallels in certain Eastern traditions such as the Neo-Confucian '*sheng*'.[4] Yet *ST* II-II.1–170, Aquinas' largest systematic account of the particular virtues, lacks a distinct treatise on the intellectual virtues, parallel with the sixth book of the *Nicomachean Ethics*, and wisdom is covered in just a single question appended to the non-Aristotelian virtue of *caritas*.[5] Furthermore, while Aquinas acknowledges that we can acquire virtues in the Aristotelian manner by repeated good actions, he also claims that proper or perfect virtues are not acquired in this manner, but *infused* in us by God.[6] These infused virtues include counterparts of many of the familiar acquired moral and intellectual virtues. For example, as well as acquired justice, there is infused justice, which is the proper sense of the virtue of justice. Similarly, besides acquired prudence, there is infused prudence, which is the proper sense of the virtue of prudence, and so on.[7] Aquinas also claims that these infused virtues differ in species and in the kind of flourishing to which they are directed, in comparison to virtues acquired in the manner of those of the *Nicomachean Ethics*.[8]

Aquinas' introduction of these infused virtues raises acute difficulties of interpretation. While the infused virtues include certain virtues that are entirely without parallel in Aristotle's texts, such as *caritas*, the claim that there are duplex homonymous moral virtues highlights, in particular, the difficulty of interpreting Aquinas' approach. How are the infused virtues to be understood? What does an infused virtue provide that is specifically different from its acquired counterpart? Why are the infused virtues the only perfect virtues? I shall refer to such questions, collectively, as the problem of the infused moral virtues.

Throughout the history of commentary, the problem of the infused moral virtues has attracted three main responses. One surprisingly common response has been simply to ignore the issue, an approach perhaps made easier by studying particular virtues in small, isolated sections, together with the fact that Aquinas rarely states explicitly whether he is referring to an infused virtue or its acquired counterpart.[9] Other commentators, more conscious of Aquinas' claims about the infused virtues, have taken the view that he is largely or wholly mistaken, arguing that the infused moral virtues are redundant.[10] A third response has been to acknowledge that the infused moral virtues are specifically different from their acquired counterparts, but to continue to treat them in broadly Aristotelian terms. An example of the latter approach can be found in the work of Réginald Garrigou-Lagrange OP, who practically defined the Neo-Thomist School during the first half of the twentieth century.[11] Garrigou-Lagrange claims that the definition of each of the four cardinal virtues is 'proportionally true' of the infused version and its acquired counterpart, with the infused virtue being 'higher' than its acquired counterpart.[12] While these claims leave open the question of the kind of proportional relationship of these duplex virtues, the metaphors that Garrigou-Lagrange puts forward to communicate the meaning of this distinction often imply some common scale. For instance, he describes acquired and infused temperance as being like two notes on a keyboard, played an octave apart, and suggests that infused justice "increases tenfold" the energies of the will compared to acquired justice (Garrigou-Lagrange 1946, 442).[13] If the term 'higher' indicates some relationship of proportional equivalence, especially in terms of some common scale, this approach suggests that the 'higher' virtue can be thought of as a proportionally greater version of its 'lower' counterpart. Since the 'lower' virtue is Aristotelian, then the infused virtue can also be understood, albeit indirectly, in Aristotelian terms. In other words, while Garrigou-Lagrange defends the fact of a distinct set of infused virtues, his use of language and choice of examples subtly re-validates a broadly Aristotelian understanding of Aquinas' work.[14]

Contemporary philosophers, however, have rendered any kind of proportional equivalence of the acquired and infused virtues inadmissible. Jeffrey Hause, for example, has drawn attention to the point that a proportional change in the good established by reason, such as greater asceticism, is insufficient to establish a difference of species (Hause 2007, 16).[15] Other philosophers have noted how many characteristics of the infused virtues are different in kind, rather than degree, from the virtues described by Aristotle. First, Jean Porter has drawn attention to Aquinas' claim that all human beings have the capacity to receive infused virtues, even those, such as children, who lack the intellectual training required for Aristotelian virtues (Porter 1992, 34).[16] Second, Porter and Bonnie Kent have observed how an infused virtue, unlike an acquired virtue, can be present with a previously acquired contrary disposition (Porter 1992, 30; Kent 1995, 728).[17] Third, Aquinas claims that the infused virtues are connected together, yet the unifying virtue is the non-Aristotelian virtue of *caritas* and Aquinas' view of the unity of acquired tendencies to do good actions is nuanced.[18] Fourth, infused virtues, unlike acquired virtues, can be infused or lost by a single action (*ST* I-II.71.4). Finally, according to Aquinas, even if a person possesses *all* the acquired virtues, if he lacks the infused virtues he cannot enter heaven.[19] Even if there is a state of flourishing corresponding to the acquired virtues alone, in Aquinas' worldview this state would not be heaven but that of Limbo, which Dante portrays in the manner of the Elysian Fields of classical paganism.[20]

Although the characteristics of the infused virtues are peculiar enough, there is another mystery that has attracted rather less attention in contemporary philosophy.[21] Consider, for instance, Aquinas' account of the virtue of courage. Following his treatise on courage as a virtue, *ST* II-II.123–138, Aquinas appends a further question describing an entirely new attribute, the *gift* (*donum*) of courage.[22] Although this gift of courage is infused, Aquinas claims that a gift is a different kind of quality from an infused virtue.[23] Even the addition of the gift of courage is not the end of the story, though, because Aquinas claims that this gift is linked to a further attribute, a *beatitude* (*beatitudo*) called "hungering and thirsting for justice."[24] This beatitude is in turn associated with two final and distinct attributes, the *fruits* (*fructus*) of patience and long-suffering (*ST* II-II.139.2 *ad* 3). So Aquinas appends a network with one gift, one beatitude and two fruits to the virtue of courage. Similar manifestations of this 'virtue-gift-beatitude-fruit' (VGBF) pattern can be found for the other cardinal and theological virtues in *ST* II-II.1–170, giving the impression of an intricate and even 'organic' structure utterly different from anything in the *Nicomachean Ethics*.[25]

A great many facts about this daunting array of infused virtues, 7 gifts, 7 beatitudes and 12 fruits can readily be distilled from Aquinas' texts. Indeed, more or less accurate summaries of these attributes and their interconnections were commonplace in old manuals of the Neo-Thomist tradition.[26] What has always been lacking from these descriptions, however, is what might be called a 'synthetic picture', an understanding that unifies the principles of these attributes, communicated by some appropriate metaphor. While the everyday experience of habituation provides an apparently satisfactory and easily understood synthetic picture of the Aristotelian virtues, no synthetic picture of the VGBF attributes has been offered in the history of commentary. Even Aquinas' most dedicated followers have largely restricted themselves to the summary and defense of his distinctions of particular attributes, rather than providing a metaphoric understanding that unifies the network as a whole.[27] This lack of a synthetic picture of Aquinas' account of virtue, in contrast to the familiar image of habituation drawn from Aristotle's texts, may be one of the reasons why Aristotelian virtue ethics, including the Aristotelian reading of Aquinas, remains so influential today.

I argue, however, that there is a synthetic picture by which Aquinas' approach can be understood, an approach that has a remarkably commonplace parallel in daily life (Pinsent 2012). As a consequence, although Aquinas' virtue ethics presupposes that there is a personal God and that the perfection of the virtues involves a relationship with God, the implications of his approach are by no means limited to a theological context. On the contrary, the work of Aquinas may help draw attention to a principle that is present in most kinds of virtue formation, but which has been obscured by the dominant Aristotelian paradigm. Clearly, a new insight into Aquinas' approach will also have a bearing on efforts to examine parallels and differences between virtue ethics in Western and Eastern philosophy, and be of interest to those involved in education and formation in general.

To identify the clues that help reveal this synthetic picture in Aquinas' text, I begin by examining those attributes that Aquinas appends to the virtues but which have been overlooked, for the most part, by recent commentators. I shall focus on the first and last of these three kinds of attributes, the gifts and the fruits of the Holy Spirit, since Aquinas claims that the gifts are foundational to the infused moral virtues, while the fruits feature as the terminating attributes of the entire network.

2 The Gifts and the Second Person

There are seven 'gifts', each of which Aquinas assigns to various theological and cardinal virtues in *ST* II-II. The cognitive gifts are understanding (*intellectus*) and knowledge (*scientia*), appended to the virtue of faith, wisdom (*sapientia*), appended to the virtue of *caritas*, and counsel (*consilium*), appended to the virtue of prudence. The appetitive gifts are fear (*timor*), appended to the virtue of hope, piety (*pietas*), appended to the virtue of justice, and courage (*fortitudo*), appended to the homonymous virtue of courage.[28] Aquinas also examines the genus of gift in *ST* I-II.68, immediately following his treatise on the genus of virtue, *ST* I-II.55–67.

ST I-II.68.2 claims that the gifts are essential to salvation, a claim that rules out the possibility that the gifts are adventitious in Aquinas' account of human flourishing. *ST* I-II.68.8 adds that, while the theological virtues are more excellent than the gifts, the gifts are more excellent than the moral and intellectual virtues, thereby taking precedence over any of the qualities described in the *Nicomachean Ethics*.[29] Aquinas even claims, in *ST* II-II.19.9 *ad* 4, that the gifts are the origins or foundations (*principia*) of the intellectual and moral virtues.

The function of the gifts is, however, more difficult to grasp than the fact of their importance. *ST* I-II.68.3 categorizes the gifts as *habitus*. This word is often translated as 'habit' in older texts, but contemporary philosophers prefer the word 'disposition' or leave the term untranslated. What is clear, however, is that the gifts are in the same category of perfective attribute as the virtues, which are also *habitus*. Furthermore, *ST* I-II.68.4 states that the gifts extend to all those things to which the virtues, both intellectual and moral, extend. Yet if the role of the gifts parallels that of the virtues, why are the gifts necessary? Aquinas' response is a principle set out in *ST* I-II.68.1:

> Now it is manifest that human virtues perfect man according as it is natural for him to be moved by his reason in his interior and exterior actions. It is fitting, therefore, for there to exist in man higher perfections, whereby he is disposed to be moved by God in a divine way. These perfections are called gifts, not only because they are infused by God, but also because by them man is disposed to be made readily moveable by divine inspiration.[30]

In this passage, Aquinas claims that what is specific about a gift is not that it is an infused *habitus*, since perfect virtues are also infused, but that, by means of a gift, a person is disposed to be moved by God. Despite the mechanistic connotations associated with the modern word 'movement', Aquinas clarifies elsewhere that gift-based movement is not coercive.[31] Furthermore, Aquinas' texts also imply that whatever is given by means of the gifts is not reducible to some propositional description of the world, a description that could, in principle, be communicated by some other means.[32] In what, then, does gift-based movement consist? Understanding this movement is the key to interpreting the gifts, which are, in turn, foundational to Aquinas' entire account of human perfection.

To answer this question, it is first important to clarify how the scenario Aquinas describes regarding the gifts differs from that of virtuous action in the more familiar Aristotelian sense. As noted previously, Aquinas holds that the gifts extend to all the things to which the virtues extend. This claim implies that anything that can be the matter of the virtues, such as, for example, food and drink, intellectual truths, and commutative relations in society, can also be the matter of the gifts. Yet in relation to these matters, the gifts dispose a person to be moved easily not by her own reason, which is the mode of operation of the virtues, but by God. In other words, instead of a dyadic person-object scenario, as in the operation of the virtues, Aquinas is describing a triadic person-God-object scenario in which one's stance towards an object is 'moved' by God, in some sense yet to be understood. An important corollary of this description of gift-based movement is that one's focus is not on the mover when being 'moved', but on whatever is the object of one's attention, such as food and drink, intellectual truths, commutative relations in society, and so on. Nevertheless, by means of the gifts, Aquinas claims that one is disposed to be moved by God with respect to these objects.

In the context of this triadic scenario, Aquinas' descriptions of the specific gifts help to shed light on what this movement means. When describing the gift of knowledge, Aquinas claims that this gift enables a "participated likeness" (*participativa similitudo*) of God's knowledge, knowledge that is absolute and simple rather than discursive, as is the case for the homonymous intellectual virtue (*ST* II-II.9.1 *ad* 1).[33] A similar notion of participation is implied by Aquinas' descriptions of the other gifts. By the gift of piety, we are moved to regard other persons as God regards them, namely as potential or actual children, and thereby our brothers and sisters.[34] By the gift of courage, we regard present dangers and potential future dangers with the kind of confidence with which God regards such difficulties (*ST* II-II.139.1).[35] By the gift of fear, we regard ourselves as God regards us, as adopted children with whom God desires to be united, and so shrink from the loss of our own good from God's perspective.[36] By the gift of counsel, we are "directed as though counseled by God," implying that we take on God's stance toward possible courses of action (*ST* II-II.52.1 *ad* 1). By the gift of understanding, we grasp what is proposed to us by God (*ST* II-II.8.4).[37] Finally, by the gift of wisdom, we are enabled to judge rightly about divine things on account of a 'connaturality' with them (*ST* II-II.45.2). In all these cases, the movement enabled by the gifts can be characterized in the following way: a triadic scenario in which a person shares in or appropriates God's stance towards some object.

Such appropriations of God's stance cannot, however, be reduced to mere disconnected imitations, as if one is merely imagining what God's stance would be like with respect to the object in question. Aquinas makes this point clear, for example, in his description of the gift of wisdom:

> Wisdom as a gift is more excellent than wisdom as an intellectual virtue, since it attains to God more intimately by a kind of union of the soul with him (*unio animae ad ipsum*). (*ST* II-II.45.3)[38]

The idea of a union or oneness of the soul with God in this passage is reflected in other texts about the gifts, especially Aquinas' claim that whoever has *caritas*, which unites us to God, has all the gifts, none of which can be possessed without *caritas* (*ST* I-II.68.5). So gift-based movement is not a mere imitation of God's stance but implies a union of the soul with God, a union based on *caritas*. This notion of interpersonal union is corroborated by a comment Aquinas makes in his commentary on the Gospel of John, when he states explicitly:

> For the Holy Spirit is the unfailing fountain from whom all gifts of grace flow, "*One and the same Spirit does all these things*" (1 Cor 12:11). If anyone has a gift of the Holy Spirit, and not the Spirit, the water is not united with its source, and so is dead, and not living.[39]

In this passage, Aquinas confirms, therefore, that gift-based movement is never possible unless the giver of these gifts is, in some sense, present to the person, a claim that is consistent with the notion that gift-based movement involves a certain union of soul with God.

Aquinas' descriptions of gift-based movement therefore express three main principles. First, this movement takes place in the context of a triadic person-God-object scenario. Second, the movement of a person with respect to the object involves a participation in God's stance towards the same object. Third, gift-based movement involves what Aquinas describes as a union or oneness of the soul with God, such that the movement is never possible in the absence of such union.

Although Aquinas is outlining a theological scenario, these descriptions of the operation of the gifts seem to parallel a remarkably commonplace interaction. This interaction is what experimental psychologists and philosophers today call "joint attention" (Eilan et al. 2005, 1). Although a precise definition of joint attention remains a matter of some debate, the phenomenon has at least the following basic characteristics. First, joint attention takes place in a person-person-object situation in which both persons share attention to some common object. In the context of parent-child interactions, activities that manifest joint attention can include, for example, referential use of eye contact, offering and giving objects to others, pointing at objects and turning to follow the pointings of the other person. Second, joint attention involves sharing a stance towards the object that is the focus of attention, a sharing that tends to involve both persons coming into alignment with one another. In gaze following, for instance, this alignment may be manifested in obvious and explicit movements, as when a child turns to follow a parent's attention to something and appropriates the parent's attitude towards the object, manifested by certain emotional cues. Third, joint attention has been described as sharing an awareness of the sharing of focus *with* the other person,[40] an experience commonly described as a 'meeting of minds.' As a corollary, one can only engage in joint attention with someone who is present, unlike a situation in which one merely imagines oneself to be in another person's situation.

0 being familiar to most people who have cared for a child, the phenomenon of joint attention attracted little scientific or philosophical investigation until the realization that there are certain human persons whose engagement in joint attention is atypical. Clara Claiborne Park, who wrote a classic book on bringing up a child with this atypical engagement, recalls the day she saw another child point to a box of candy. She suddenly realized, "I had never seen Elly [her own child] point." Reflecting on the significance of what was missing in Elly's behavior, she remarks:

> To point is so simple, so spontaneous, so primary an action that it seems ridiculous to analyze it. All babies point, do they not? To stretch out the arm and the finger is, symbolically and literally, to stretch out the self into the world—in order to remark on an object, to call it to another's attention, perhaps to want it for oneself. From pointing comes the question "What's that?" that unlocks the varied world. To point, to reach, to stretch, to grab is to make a relation between oneself and the outside. To need is to relate. (Park 1967, 6)[41]

In this passage, Park observes how the gesture of pointing is linked to relations with another person, since pointing calls an object to another's attention. Conversely, the failure to engage in such activities is associated with a failure to relate to persons in this way, as in the case of Elly. The kind of pointing that Park describes is what would today be classified as an activity of joint attention. Given Elly's lack of engagement in such activities, it is not surprising that Park also remarks on Elly's inability to relate to persons in general, sometimes describing her as not 'seeing' persons at all (Park 1967, 93).

Such behavior falls under the set of typical characteristics of what is today called 'autism' or, more correctly, 'autistic spectrum disorder', a condition that has been both a means and a motive to investigate joint attention scientifically. Although Park describes Elly as not 'seeing' persons, based on her atypical reactions to other people, recent empirical studies suggest that those with autism are not unusual in their ability to distinguish persons from other kinds of beings in the world.[42] What is unusual, however, is the mode of engagement that they have with other persons. In the following passage, Peter Hobson, a professor of developmental psychopathology, describes the outcome of a study designed to highlight how autistic children interact with other persons:

> In two respects, then, the children with autism were not moved to adopt the orientation of the person they were watching. They did not adopt the style with which the experimenter executed the actions, nor did they identify with him and copy his self-orientated actions so that these actions became orientated towards themselves. On the other hand, they were perfectly able to perceive and copy the strategies by which he achieved the goals in each demonstration. So they were able to learn something from watching what the experimenter did … Yet what they learned seemed to be available from their position as a kind of detached observer of actions and goals. They were not 'moved'. (Hobson in Eilan et al. 2005, 200)

The experiment described in this passage showed that autistic children were perfectly capable of recognizing other persons and following instructions given by another person.[43] Furthermore, they were also able to imitate the actions and goals of another. What the children with autism did not do, however, was to appropriate another person's psychological orientation or to 'identify' with the other person.[44] Such language has clear parallels with the notion of a participation in God's stance and a union of soul with God in Aquinas' descriptions of gift-based movement. Furthermore, Hobson even articulates what is missing from autistic behavior by the same language that Aquinas uses to describe the gifts: the children with autism were not 'moved'.

Autism may therefore help to provide a metaphoric understanding of the role of the gifts in Aquinas' virtue ethics, namely that the gifts remove a person's 'spiritual autism' in relating to God. In making such a claim, I do not in any way conclude or intend to imply that the condition of autism inhibits a person from receiving the kinds of perfective attributes to which Aquinas refers. On the contrary, in contrast to the rather elitist virtue ethics of Aristotle, Aquinas claims that anyone can receive the infused gifts (Cf. *ST* II-II.47.14 *ad* 3). Furthermore, the capacity to be moved by non-divine persons is not always an unqualified benefit. One cannot, for example, imagine someone with severe autism to be moved (except perhaps to distress) by the hysterical reactions of a crowd cheering a dictator. Nevertheless, autism does have a pedagogical role, insofar as this condition seems to offer a suitable metaphor for what Aquinas considers a human being—*any* human being—to be like without the perfective attributes he describes, with respect to the one who infuses these attributes.

The parallel with autism also provides another insight. As Leo Kanner noted in his original description of the syndrome, and many subsequent studies have confirmed, children with autism often refer to themselves as 'you' and the person they are speaking with as 'I' (Kanner 1968). For example, the question, "How are you?"

might elicit the response, "You are fine," a response that would be correct if the term 'you' denoted some third referent, as in the question and response, '"How is she?" ... "She is fine".' In other words, the autistic child will often instinctively use a second-person pronoun in the manner of a third-person pronoun. This difficulty in grasping the proper use of the second-person mode of address is consistent with a lack of familiarity with joint attention, given that 'you' is the natural way of addressing a person with whom one is engaging in joint attention, and that joint attention and the proper use of 'you' share the characteristic of pertaining only to situations of mutual personal presence.[45] Further evidence for this connection is provided by the fact that children of comparable linguistic ability with non-autistic retarded learning conditions, such as Down Syndrome, do not normally reverse the typical use of 'I' and 'you' pronouns in conversation.[46]

This correlation between autism and the difficulty in grasping the 'you' mode of address implies another way of characterizing the kind of relatedness enabled by joint attention.[47] Joint attention enables what one could describe as *second-personal* relatedness with another human being, a mode of relating to a particular person who is present to oneself and properly designated by the deictic term 'you'.[48] Insofar as joint attention also provides a metaphoric understanding of gift-based movement in Aquinas' account of human flourishing, this characteristic suggests that the gifts enable specifically *second-personal* relatedness with God. If, therefore, virtues can be classified as *first-personal habitus*, by which a person is moved easily by her own reason, the gifts are the associated *second-personal habitus*, by which a person is moved by God in the manner of joint attention. Such a conclusion is consistent with the intrinsic connection that Aquinas makes between the gifts and *caritas*, given that Aquinas describes *caritas*, the supreme theological virtue, as friendship with God and friendship implies the ability to relate to one's friend as to a specific second person. (*ST* II-II.23.1)[49]

3 The Fruits and Interpersonal Resonance

Aquinas introduces the notion of a 'fruit' by means of a natural metaphor. Using the example of a plant, Aquinas observes that a material fruit is "the product of a plant when it comes to perfection, and has a certain sweetness." (*ST* I-II.70.1) He claims that these qualities also characterize the ethical attribute which he calls a 'fruit':

> The notion of fruit implies two things: first that it should come last; second, that it should calm the appetite with a certain sweetness and delight ... that which is last simply, and in which one delights as in the last end, is properly called a fruit; and this it is that one is properly said to enjoy. (*ST* I-II.11.3)[50]

Consistent with the notion of a fruit coming last, Aquinas generally assigns the fruits of each virtue last in the *ST* II-II.1–170, after the appended gifts and beatitudes. Aquinas also examines the genus of fruit in *ST* I-II.70, after examining the genera of gift and beatitude.

What, however, does Aquinas understand by the term 'fruit'? According to *ST* I-II.70.3, the fruits of *caritas*, joy (*gaudium*), peace (*pax*), patience (*patientia*) and long-suffering (*longanimitas*) pertain to the good ordering of the mind. The fruits of goodness (*bonitas*), benignity (*benignitas*), meekness (*mansuetudo*) and faith (*fides*) pertain to one's neighbor. The fruits of modesty (*modestia*), continency (*continentia*) and chastity (*castitas*) pertain to the body. The 12 fruits therefore extend to many, and perhaps all, of the same things to which the virtues and the gifts extend. Why, therefore, is there yet another set of attributes, distinct from the virtues and the gifts?

The mystery of the fruits is deepened by Aquinas' claim that they are not *habitus* in the manner of gifts or virtues.[51] On the contrary, Aquinas describes the fruits variously as *actus* or *operationes* or *opera* of the virtues and as the last and congruous products of the gifts.[52] The implication is that a fruit is some kind of actualization or manifestation of the virtues and gifts. Some of the fruits, such as *caritas*, have homonymous counterparts among the virtues, whereas others, such as joy and peace, are unique to the category of fruit. How, then, is the attribute of fruit to be understood?

Aquinas implies an answer to this question in his description of the fruit of peace, which he classifies as an *actus* of *caritas*:

> Peace implies a twofold union ... The first is the result of one's own appetites being ordered toward one thing; while the other results from one's own appetite being united with the appetite of another: and each of these unions is effected by *caritas* ... hence it is reckoned a sign of friendship if people '*make choice of the same things*' (*Ethic*. ix, 4), and Tully says (*De Amicitia*) that '*friends like and dislike the same things*.' (*ST* II-II.29.3)[53]

In the article cited above, Aquinas highlights the interpersonal aspect of the fruit of peace and its connection with friendship, observing that peace implies a harmony between two persons, an alignment in which one chooses the same things as one's friend and with one's friend.[54] Aquinas also describes other fruits in interpersonal terms: he associates joy with the experience of God abiding in a person, and the fruit of *caritas* with both a union of affections and a union of persons (*ST* II-II.28.1; *ST* II-II.27.2).

These interpersonal descriptions are consistent with Aquinas' claim that the fruits are the last and congruous products of the gifts. Since, as argued previously, the gifts are second-personal *habitus*, one would expect the fruits, as *actus* of these gifts, to have a second-person aspect as well. Nevertheless, in Aquinas' network of perfecting attributes, not all *actus* of the gifts are fruits. The *actus* of counsel, for example, are not fruits, but only means for bringing about fruits.[55] Furthermore, the fruits are also consequent upon the beatitudes, as Aquinas states in *ST* II-II.139.2,[56] implying some intermediate step between gift-based *actus* in general and the special class of *actus* that are fruits. By what measure, then, are the fruits perfect or complete compared to gift-based *actus* in general?

Some poetic inspiration to help answer this question may be found in a scene from Dante's *Paradiso*, during which Dante the traveler encounters the soul of Aquinas in heaven:

> And he will have some shadowing forth of that
> True constellation and the double dance
> That circled round the point at which I was;

> Because it is as much beyond our wont,
> As swifter than the motion of the Chiana
> Moveth the heaven that all the rest outspeeds.
> (*Paradiso*, Canto XIII, 19–24)[57]

The idea of 'stars', that is, souls in paradise, revealing their interpersonal union to Dante by the image of a perfect dance, suggests that it is important to examine not only what it means to be moved by another person, but the manner in which such movements are perfected. Orchestras, choirs, or dancers, for example, do not always harmonize very well and can experience discords, mistiming, and mistakes. When their joint operation becomes second nature, however, with the participants being near-perfectly attuned to one another, there is often a sudden, disproportionate improvement in the objective quality of the activity. Subjectively, musicians or dancers may also experience the pleasurable exhilaration of 'flying along', accompanied by the sense of being 'one' with the other persons. This harmonized operation appears similar to the phenomenon of *resonance* in physics, when two systems engage in close to perfect joint operation with disproportionate ease and intensity.[58] In the case of the joint operation of two persons, when activities in union with the other person have become second nature, the phenomenon could be described as an *interpersonal resonance*.

Can the fruits be regarded as resonances? A resonance is a perfected joint operation, matching the description of the fruits as *actus* that are also perfections, the terminating attributes of the entire network of gifts, virtues and beatitudes. An interpersonal resonance also needs at least two persons, consistent with the inference that the fruits, insofar as they are based on the gifts, are based on a form of second-person relatedness between persons. A resonance in joint human activity, such as musicians playing in perfect harmony, is also pleasurable, matching Aquinas' description of a fruit as having 'sweetness.' Finally, in the context of such activity, a resonance indicates that some aspect of another person is sealed or imprinted on oneself, a mode of knowing another person which is not mediated but immediate.[59] Within the context of a harmonious piece of music, for example, the musicians can be said to have a limited but nevertheless immediate sense of interpersonal union, a description that may parallel the way in which Aquinas associates the fruits with having an immediate experience of God being with oneself or even abiding in oneself (*ST* II-II.29.3; *ST* I-II.65.5). I therefore suggest that the fruits, as Aquinas describes them, are various modes of a kind of perfectly realized second-person relatedness, an interaction that can be characterized as a resonance.

4 Implications of the Second Person for Virtue Formation

If second-person relatedness, established in situations of joint attention between human persons, provides a key to Aquinas' descriptions of the gifts and the fruits, the same metaphoric understanding, if correct, should also help to interpret the peculiar characteristics of the infused virtues. As noted previously, the goal is to obtain a synthetic picture that serves to unify Aquinas' multifarious, extremely complex account of human perfection. If certain stated characteristics of the infused virtues are incompatible with second-person relatedness, then this type of interaction would seem implausible as a successful candidate for this required synthetic picture.

A preliminary indication that second-person relatedness does have explanatory power for Aquinas' descriptions of the infused virtues is the special role assigned to *caritas*. As has been noted, *caritas* is the first of the named fruits. The homonymous virtue of *caritas* plays a role in Aquinas' account of virtue that is analogous to the role of prudence in the *Nicomachean Ethics*. Just as prudence unifies the moral virtues in the Aristotelian account, *caritas* unifies the entire system of infused virtues and gifts. Aquinas claims that a person with *caritas* has all the infused virtues and gifts; conversely, without *caritas*, a person has none of these *habitus*. Since Aquinas defines *caritas* as friendship, and since friendship is a fruition of second-person relatedness in everyday life, the idea that second-person relatedness has explanatory power across Aquinas' system of infused *habitus* is certainly plausible.

Furthermore, the second-person aspect is clear in the case of the gifts and shows that Aquinas' claim that *caritas* unites the gifts is not merely a matter of words or an assertion about some kind of extrinsic association. Gift-based movement involves a person-person-object scenario in which one person appropriates the stance of the other personal agent in a certain 'union of soul.' These two characteristics of gift-based movement are also proper to friendship since, as noted previously, "friends like and dislike the same things," and one cannot be friends with someone if one has no desire for mutual presence and union with one's friend. So the gifts imply a disposition towards friendship. Conversely, the absence of any disposition towards friendship suggests that gift-based movement would also be inhibited. Since Aquinas defines *caritas* as friendship, and a disposition towards friendship characterizes the gifts, then *caritas* and the gifts are inherently connected.

In the case of the infused moral virtues, the connection with the second-person relatedness is slightly more obscure than in the case of the gifts. In the text on the main principle of the gifts, *ST* I-II.68.1, cited previously, Aquinas claims that "human virtues perfect man according as it is natural for him to be moved by his reason in his interior and exterior actions." In other words, a human virtue in general disposes a person to move himself. How, then, can a virtue in general, which is a first-personal *habitus*, have an inherent connection to second-person relatedness?

Certain indications of such a connection can be found among the many subtle changes of emphasis in Aquinas' treatment of the moral virtues in *ST* II-II.47–170 compared to equivalent sections of the *Nicomachean Ethics*. One example is the way Aquinas introduces novel virtues and actions of virtues that are closely connected with the principle of *caritas*, such as humility and prayer.[60] Similarly, Aquinas highlights the seriousness of certain vices that would appear to be especially damaging to *caritas*, such as reviling, back-biting, tale-bearing, derision, or cursing, assigning a prominence to these vices that is not found in the *Nicomachean Ethics* (*ST* II-II.72–76). Nevertheless, even though many of the specific virtues that Aquinas describes are compatible with the notion of second-person relatedness, the challenge remains of accounting for the non-Aristotelian characteristics of the virtues in general. In what sense, in particular, can a virtue ever be said to be gained or lost immediately, as opposed to strengthened or diminished gradually through habituation? This immediacy is so counterintuitive, in comparison with the classical understanding of virtues like temperance, that it scarcely seems appropriate to classify such *habitus* as virtues at all.[61]

Upon closer examination, however, even this peculiar immediacy of infusion or loss of the infused virtues may be explicable in terms of second-person relatedness. What Aquinas actually says is that when *caritas* is banished, by one seriously evil act, all the infused virtues are "expelled", "excluded" or "cut off" (*excluduntur*) (*ST* I-II.71.4). In other words, such exclusion does not mean that all habitual dispositions previously associated with good actions suddenly vanish, but that any such dispositions that remain cease to be effective as virtues. An example from human relationships may help to shed light on what Aquinas might mean by this claim. Consider, for example, a couple who have been happily married for many years. If one of the spouses suddenly betrays the other, the spouse who does the betraying will not suddenly lose all the good habitual dispositions of daily life. Nevertheless, dispositions to eat and drink moderately, to clean the house regularly and so on will not be conducive to the flourishing of the relationship with the other person so long as the relationship has been betrayed and there has been no act of reconciliation. Conversely, if and when an act of reconciliation is carried out, then the good dispositions of daily life will once again be re-ordered or re-connected with the flourishing of the relationship.[62] This example illustrates how dispositions can indeed be immediately lost or gained, at least as virtues with respect to the flourishing of a relationship, insofar as the relationship itself can be betrayed or reconciled by means of a single action. This characteristic of second-person relatedness, ordered towards friendship, is not only compatible with Aquinas' claim about the immediate infusion or loss of the infused virtues, but is a surprisingly straightforward way of understanding this claim.

Furthermore, the principle of second-person relatedness goes further than showing a new way of understanding Aquinas' claims about the virtues. The introduction of this principle also raises the question of whether the Aristotelian paradigm has caused important aspects of everyday virtue formation to be overlooked. Even in the case of temperance, the archetypal Aristotelian moral virtue, habituation is only part of the story even in everyday life. Consider the case of an infant learning to eat the right amount of food at the right time. Infants often have a lack of interest in eating what they should, when they should, and are often far more interested in the food belonging to their parents or caregivers than the food that is set before them. To encourage the child to eat, a parent will often have to play a game with the infant, in which the infant takes delight in the activity with the parent rather than the food in itself. In other words, joint attention activities, manifesting second-person relatedness, are surprisingly common even in promoting the early acquisition of temperance in everyday life.

These examples of virtues in the context of marriage and feeding a child also anticipate and respond to a possible objection to the value of Aquinas' work on the virtues. As noted previously, Aquinas' treatment of the virtues is explicitly theological, in the sense that he presupposes that there is a God, that God is personal, that God desires friendship with human beings and provides the capacities required for human beings to enjoy such a relationship. Consequently, these premises might imply that Aquinas' theological virtue ethics has little or no interest to those who do not share his theological worldview. Yet the terms and metaphors that Aquinas uses to describe a relationship to God are themselves drawn from relationships among human persons. Aquinas' insights into the interoperation of the infused virtues and gifts may therefore highlight the way in which second-person aspects of everyday virtue formation have tended to be neglected in the light of the Aristotelian paradigm. Furthermore, subtle empirical tests have already shown ways in which the descriptions of affective states and emotional responses given by those with autism are often atypical, especially in situations involving some appropriation of another person's stance.[63] In future, it may be possible to extend such work to study how autism modifies character formation, helping to put the study of the impact of second-person relatedness on virtue formation on an empirical basis.

Finally, the question may be asked whether the role of second-person relatedness in Aquinas' account of character formation might have parallels in certain schools of Eastern philosophy. Certainly, the notion of virtue as a disposition arising from a fine internal state is a recognized characteristic of human flourishing in Eastern philosophy.[64] Furthermore, the notion of harmony (*he*) is an ancient value in China and lies at the heart of Neo-Confucianism, a characteristic that seems similar, at first glance, to resonance as a metaphoric understanding of the fruits in Aquinas' thought.[65] The challenge, I believe, will be to establish whether the principle of harmony in Eastern thought has a specific second-person basis, disposed towards friendship, in contrast to a more generic goal of coherence among all things. Since, however, the process of establishing the specificity and characteristics of second-person relatedness is still a comparatively new field of research, the work of comparison with Eastern philosophy may have to await further progress in analytic philosophy, psychology, and related fields.

In conclusion, while an Aristotelian interpretation of Aquinas' intricate descriptions of the virtues and associated attributes has become implausible, recent work on joint attention has now suggested a new synthetic picture based on second-person relatedness. According to this interpretation, the goal of the various infused *habitus* in Aquinas' account is to remove a commonly experienced spiritual autism with respect to God, enabling second-person relatedness with God, the fruition of which is friendship. Beyond Aquinas' theological context, his work may help to highlight second-person influences on virtue formation in general, providing an alternative to the dominant Aristotelian paradigm and future topics for comparison with aspects of Eastern philosophy and education in general.[66]

End Notes

1. The treatise on the virtues, vices and associated matters in general, *ST* I-II.55–89, has 189 articles; the treatise on the particular virtues and associated matters, *ST* II-II.1–170, has 815 articles.
2. References to the virtues and associated matters are found in numerous places besides the *ST* and In NE. Key sources include *Scriptum super libros Sententiarum* (In Sent), III, d.33–35, *Quaestiones disputatae de virtutibus in communi* (QDVC) and *Quaestiones disputatae de caritate* (QDC). Within the Scriptural commentaries, *Reportatio super Epistolam ad Galatas* (In Gal), 5, is especially important for presenting the integration of the virtues with other kinds of perfective attributes.
3. A clear sign of Aristotelian influence can be found in Aquinas' definitions of the matter of each virtue, in other words, the kinds of things in the world to which each virtue relates. In *ST* II-II.141.4, 5, for example, Aquinas follows Aristotle in defining the matter of the virtue of temperance as the desires and pleasures of touch. Similarly, in every other instance in which Aquinas introduces a virtue with a homonymous counterpart in the work of Aristotle, he adopts Aristotle's definition of the matter of the virtue explicitly: prudence in *ST* II-II.47.2, 3, 8, 9, 10, 12; justice in q.58, a.1, 5, 6, 8; courage in q.123, a.3; liberality in q.117, a.2; magnanimity in q.129, a.2 and magnificence in q.134, a.3. The fact that this list includes certain virtues, such as magnanimity, that initially seem at variance with the principles of Aquinas' own religious tradition, gives additional credence to an Aristotelian reading.
4. See, for example, Angle (2009, 13–29).
5. *ST* II-II.45, with vices opposed to wisdom treated in q.46. Furthermore, the kind of wisdom described in these questions is not a virtue, but a different kind of perfective attribute appended to *caritas*, a word sometimes translated as 'love' or 'charity'. Since, in Aquinas' work, *caritas* denotes the unique love pertaining to divine friendship (*ST* II-II.23.1), I have left the word untranslated.
6. *ST* I-II.65.2 c, "Solae virtutes infusae sunt perfectae, et simpliciter dicendae virtutes, quia bene ordinant hominem ad finem ultimum simpliciter." According to Aquinas, only the acquired virtues are caused in an Aristotelian manner, that is, by repeated good actions, whereas infused virtues, that is, virtues in the proper or perfect sense, are caused by the action of God. See, for example, QDVC, a.2, ad 18, in which Aquinas claims that, just as acquired virtues are increased and fostered by the same sort of acts which caused them, so the infused virtues are increased by the action of God, by whom they are caused.
7. Aquinas differentiates acquired and infused justice in *ST* I-II.100.12, claiming that only the latter is true justice. In *ST* I-II.47.14, he distinguishes acquired and infused prudence. In *ST* I-II.63.4 he describes acquired and infused temperance as distinct species of temperance. Acquired and infused courage are mentioned

as distinct virtues in QDVC, a.10, *ad* 10. John Inglis has drawn attention to Aquinas' introduction of infused counterparts of the acquired moral virtues and to the need for greater study of the impact of infused virtues on Aquinas' ethics as a whole (Inglis 1999). Robert C. Miner has examined prudence in *ST* I-II, and argued that Aquinas' descriptions differ strikingly from Aristotelian *phronêsis* (Miner 2000). Most of the intellectual virtues are, however, replicated as gifts rather than virtues (*ST* II-II.8, 9, 45), a point made by Eleonore Stump (Stump 1999, 48).

8. An explicit statement that the acquired and infused virtues differ in species can be found, for example, in In III Sent., d.33, q.1, a.2, qc. 4, co.: "Virtutes acquisitae et infusae differunt specie." The fact that the infused virtues are directed to a specific kind of flourishing is highlighted in *ST* I-II.63, which refers to an end that is 'supernatural' (*supernaturalis*), an end to which infused virtues are ordered and which no acquired virtues alone can attain (cf. *ST* I-II.63.3 *ad* 3, QDVC, q.1, a. 10). The term 'supernatural', however, despite its connotations in popular culture, merely means 'above' or 'beyond' nature. In other words, considered in isolation the word is negative insofar as its referent is specified by that which it is not, rather than that which it is. It is the study of the infused virtues and associated matters that may shed light on what the term 'supernatural' denotes in a positive sense in Aquinas' thought.
9. In the late Middle Ages, the study of virtue ethics was often synonymous with holding lectures and disputing questions on the issues raised by the *Nicomachean Ethics*, an approach that encouraged Aquinas' texts to be consulted in a fragmentary way and the issue of the infused virtues to fade into the background of scholarly concern. See Flüeler (2007, 277). To cite a recent example, Robert Miner's work excels at correcting many misconceptions regarding Aquinas' account of the passions and is, in many ways, an outstanding work of scholarship (Miner 2009). Yet Miner does not address directly the question of the species of the virtues, infused or acquired, that perfect the passions of *ST* I-II.22–48, despite the importance of clarifying this issue. In other words, in this instance as in many other commentaries, past and present, the problem of the infused virtues has been ignored.
10. As Bonnie Kent has noted, Duns Scotus, for example, argued a principle of economy against Aquinas' approach. According to Duns Scotus, only seven virtues are required to make a human being perfect: the theological virtues of faith, hope and *caritas*, together with the naturally acquired moral virtues of justice, temperance and courage and the naturally acquired intellectual virtue of prudence. See Kent (2002, 354), cf. Scotus, *Ordinatio* 3, suppl. d. 36.
11. Réginald Garrigou-Lagrange OP exerted an unparalleled influence as a teacher at the Pontifical University of St. Thomas Aquinas, known as the Angelicum, in Rome from 1909 to 1960.
12. See Garrigou-Lagrange (1946). When describing prudence, for example, he states "Cette définition est proportionnellement vraie de la prudence acquise, éclairée par la lumière naturelle de la raison, et de la prudence infuse, éclairée par la lumière infuse de la foi," (Garrigou-Lagrange 1946, 529). When describing

courage, he also refers to the same definition being 'proportionally true' of both acquired and infused courage, "Cette définition est vraie proportionnellement de la force acquise du soldat qui expose sa vie pour la défense de sa patrie, et de la force infuse qui, sous la direction de la foi et de la prudence chrétienne, reste ferme malgré toutes les menaces dans la voie du salut, comme on le voit chez les martyrs," (Garrigou-Lagrange 1946, 536). In other words, Garrigou-Lagrange conflates the definitions of the acquired and infused versions of the cardinal virtues, holding that the same definition holds proportionally for both species.

13. "Il y a ainsi une *différence spécifique* entre la tempérance acquise et la tempérance infuse, différence analogue à celle d'une octave, entre deux notes musicales de même nom, séparées par une gamme complète" (Garrigou-Lagrange 1946, 442). "Elles décuplent les énergies de la volonté." (Garrigou-Lagrange 1946, 534).
14. One mark of the enduring influence of this understanding of the infused virtues is that theologians trained in the Neo-Thomist tradition sometimes persist in employing the metaphor of height even when arguing against details of the Neo-Thomist interpretation. For example, an appendix to the 1964 translation of the ST, edited by Thomas Gilby O.P, claims that, "The supernatural does not derogate from the natural, but witnesses to our human dignity, for if impotent of ourselves *to scale the heights* [my italics] our impulse is towards them." In other words, the flourishing to which the infused virtues are ordered is presented in this text in terms of heights that can be seen but not scaled, an expression of the natural desire for the supernatural which Garrigou-Lagrange would reject. Nevertheless, the implied difference between the capacities of the acquired and infused virtues is still expressed in terms of a vertical displacement. See Aquinas (1964, 101).
15. Hause's argument is important because Aquinas' example of infused temperance (*ST* I-II.63.4) seems to suggest that the principal difference between acquired and infused temperance is simply that the latter is more ascetic, promoting fasting and mortification rather than moderate consumption. From this example, it might be thought that the infused virtues are proportionally harder or more demanding versions of the acquired virtues, as implied in the commentary of Garrigou-Lagrange (1946, 449). Hause, however, argues that such an approach is insufficient to account for the distinction, since acquired temperance could also be constituted in a mean that pursues fasting and chastising. Infused and acquired virtues must be different in kind, not merely in degree, and it is insufficient merely to state that they are different in kind (as Garrigou-Lagrange does) without also accounting for this difference.
16. Citing *ST* II-II.47.14, Porter points out that infused prudence is present, according to Aquinas, in all who have grace, even those who cannot exercise independent thought and judgment.
17. The article from which these observations are taken is *ST* I-II.65.3, *ad* 2, "Similiter habitus moralium virtutum infusarum patiuntur interdum difficultatem in operando, propter aliquas dispositiones contrarias ex praecedentibus actibus

relictas. Quae quidem difficultas non ita accidit in virtutibus moralibus acquisitis, quia per exercitium actuum, quo acquiruntur, tolluntur etiam contrariae dispositiones." See also *ibid. ad* 3, "Aliqui sancti dicuntur aliquas virtutes non habere, inquantum patiuntur difficultatem in actibus earum, ratione iam dicta; quamvis habitus omnium virtutum habeant."

18. Aquinas is often held to defend a strong thesis of the unity of the virtues, a position that opens him to the kind of criticism that MacIntyre makes with his example of the courageous Nazi (MacIntyre 2007, 179–180). Yet Mark Jordan points out that Aquinas' view of the unity of the virtues is nuanced (Jordan 1993, 240). Aquinas does hold that the infused virtues are unified, being infused together with *caritas* and connected on account of *caritas* (*ST* I-II.65.3), but his position regarding the unity of the acquired virtues is more ambiguous. In fact, if an acquired moral virtue is regarded simply as an acquired tendency to carry out a particular kind of good action, then Aquinas argues that such virtues are not connected, "since we find men who, by natural temperament or by being accustomed, are prompt in doing deeds of liberality, but are not prompt in doing deeds of chastity." (*ST* I-II.65.1).
19. Aquinas implies that possession of the acquired virtues, without the infused virtues, could even be a worse state than being without any of the virtues, based on the following argument. Acquired virtues imply moral and intellectual maturity, yet once a person has achieved such maturity, she will tend, among other things, to deliberate about herself. If this deliberation leads her to direct herself to her due end, she will, by means of grace, receive the remission of original sin, together with *caritas* and the infused virtues (*ST* I-II.89.6). If, however, a person does not direct himself to the due end, and as far as he is capable of discretion at that particular age, he will sin mortally, for through not doing that which is in his power to do (*ST* I-II.89.6) and if unrepentant, he will end in hell (*ST* I-II.89.6; QDM, q.7, a.10, *ad* 10). Similarly, if he accepts grace, and subsequently rejects it by mortal sin, he will again forfeit the infused virtues, without necessarily losing the acquired virtues. Once again, if he persists in mortal sin without repentance, he will end in hell. So to be in a state of possessing only the acquired virtues is of no true benefit to the moral agent. Indeed, to possess acquired virtues without infused virtues may even be a sign of the rejection of grace arising from some conscious decision, such as the decision to commit a seriously evil action, arguably a worse state than one in which a person lacks the maturity to choose.
20. Dante, *The Inferno*, Canto IV.
21. An exception, noted previously, is Stump (1999, 48). Stump draws attention to the fact that wisdom, understanding and knowledge are described as gifts, not virtues, in *ST* II-II, specifically in qq. 8, 9 and 45 respectively. Another exception is Rebecca Konyndyk De Young, who draws attention, albeit in a brief reference, to the existence of the gifts and fruits (DeYoung 2003, 149). Such observations, however, only describe certain features or impressions of the structure of virtues, gifts, beatitudes and fruits. Contemporary philosophers have not, as yet, made any systematic study of the structure as a whole.

22. Aquinas introduces the gift of courage in *ST* II-II.139.1, at the conclusion of his treatment of the virtue of courage.
23. Aquinas devotes *ST* I-II.68.1, I-II.69.1 and I-II.70.2 to arguments that the virtues, gifts, beatitudes and fruits are distinct kinds of attributes from one another.
24. Aquinas argues that courage is about difficult things and, since it is difficult to do virtuous deeds with an insatiable desire, which may be signified by 'hunger and thirst for justice', this beatitude is properly assigned to the virtue of courage (*ST* II-II.139.2) as well as to the virtue of justice (*ST* II-II.121.2).
25. Aquinas' network is described as an 'organic unity' in Pinckaers (2003, 87). By contrast, most commentators on the non-Aristotelian aspects of Aquinas' network of attributes have tended to concentrate on the gifts, paying little attention to the virtue-gift-beatitude-fruit network as an organic whole. The pattern of this approach was set at least by the seventeenth century in a classic work by João Poinsot (John of St. Thomas), *De donis Spiritus sancti*, from volume 5, d.18 of his *Cursus theologicus*. As the title implies, John of St Thomas concentrates almost exclusively on the gifts, devoting 10 out of 11 chapters to commentaries on their general or particular operations. While he makes occasional references to the beatitudes and fruits in these chapters, it is only in chapter IX that he considers the beatitudes as a distinct set of attributes and it is only in the last two paragraphs that he mentions the fruits. For an English translation of this work, see John of St. Thomas, *The Gifts of the Holy Ghost*, trans. Dominic Hughes O.P. (New York: Sheed & Ward, 1951). While John of St Thomas did intend to extend his work to study the beatitudes and fruits in more detail, this plan was never carried out as d.18 was the last *disputatio* he wrote in the series.
26. See, for example, Tanquerey (1930, 609–637).
27. A paradigm of this approach from the fifteenth century is that of John Capreolus, known as the 'Prince of Thomists'. His *Defensiones Theologiae Divi Thomae Aquinatis* is organized as a commentary on Peter Lombard's *Sentences* in which Capreolus collates relevant texts from Aquinas' work in response to adversaries such as Durandus of Saint Pourçain and Duns Scotus. An example of this approach from the twentieth century is that of the Neo-Thomist manuals mentioned previously.
28. The questions devoted to the specific gifts are *ST* II-II.8 (understanding), q.9 (knowledge), q.19 (fear), q.45 (wisdom), q.52 (counsel), q.121 (piety), q.139 (courage).
29. The statement that the gifts are more excellent than the moral or intellectual virtues is ambiguous insofar as this statement does not qualify these virtues as infused or acquired. Nevertheless, since Aquinas also claims that infused virtues are more excellent than any virtues acquired in an Aristotelian manner, the implication, in either interpretation, is that the gifts take precedence over any of the qualities described in the *Nicomachean Ethics*.
30. *ST* I-II.68.1, "Manifestum est autem quod virtutes humanae perficiunt hominem secundum quod homo natus est moveri per rationem in his quae interius vel exterius agit. Oportet igitur inesse homini altiores perfectiones, secundum

quas sit dispositus ad hoc quod divinitus moveatur. Et istae perfectiones vocantur dona, non solum quia infunduntur a Deo; sed quia secundum ea homo disponitur ut efficiatur prompte mobilis ab inspiratione divina." When translating the Latin, I have found the second and revised translation in 1920 by the Fathers of the English Dominican Province helpful; for a modern edition, see *The Summa Theologica of St. Thomas Aquinas* (Aquinas 1981). I have, however, preferred to use my own translations in many cases and am grateful to Fr Kevin Flannery SJ for the opportunity to discuss the most appropriate way of expressing certain ideas in Aquinas' texts. Descriptions of the gifts as enabling us to be moved by God can also be found in *ST* I-II.68.4; II-II.52.1; II-II.52.3; III.7.5.

31. Aquinas argues in *ST* I-II.10.4 that the will is not moved of necessity by God. In whatever manner, therefore, God does cause our acts of will (and this is a point of some debate), it is generally acknowledged that, according to Aquinas, God's action on the will is 'not coercive'. See Shanley (1998, 113).
32. In *ST* II-II.8.5, Aquinas refers to the Holy Spirit enlightening the mind without the gifts, in regard to truths that are preambles of faith. Nevertheless, he also states that such enlightenment, no matter how extensive, is no substitute for movement by means of the gift of understanding. He confirms this point in *ST* II-II.9.3 *ad* 3, in which he says that not everyone who understands, has the gift of understanding, but only the one who understands from a *habitus* of grace (cf. *ST* II-II.9.3 *ad* 3 on knowledge). In other words, Aquinas allows for the possibility that two individuals could genuinely understand some revealed truth, while only one of them has the gift of understanding. What this claim appears to imply is that whatever is communicated by means of the gifts cannot be reduced to propositions alone, since whatever is communicated by propositions does not depend on the means of communication. If this reduction were possible, then what is communicated by the gifts could in principle be communicated some other way, such as by tablets of stone.
33. Note that none of the cognitive gifts are deliberative, despite the fact that three of them share names with intellectual virtues that pertain to discursive reasoning. The operations of the cognitive gifts are, therefore, consistent with understanding them as enabling participation in God's absolute, 'simple' cognition.
34. In *ST* II-II.121.1, Aquinas describes the gift of piety as moving us to have a filial affection toward God, and, as a consequence, to pay veneration and service to all people on account of their relationship to God.
35. The gift of courage could be said to 'cement' the virtue of courage, because the virtue adequate for current difficulties might still be feared inadequate for hypothetical future difficulties. The gift, however, enables a sharing in God's standpoint, a confidence that any future challenge can also be overcome with divine assistance.
36. In *ST* II-II.19.9, Aquinas describes the gift of fear as moving us to revere God and avoid separating ourselves from him. This gift enables 'filial fear', the kind of fear a child has to disappoint a father whom he loves, not the fear of loss or pain for oneself, which is 'servile fear' (*ST* II-II.19.5). So the gift of fear disposes one to shrink from separation from God, not from the point of view of

one's own good as desired by oneself, but from the point of view of one's own good as desired by God with whom one is united by *caritas*.

37. The relational aspect of the gift of understanding is a point of distinction between the gift and the homonymous virtue. In *ST* II-II.8.4, Aquinas illustrates the operation of the gift of understanding by reference to John 8:12, "*He that follows me, walks not in darkness*." In II-II.8.2 he cites an incident in Luke's Gospel (Lk 24:27, 32) where Christ is described as opening the scriptures to His disciples, that they might understand them. These statements are significant insofar as they communicate the kinds of operations that Aquinas believes appropriate for illustrating this gift. These illustrations imply that there is an inherently interpersonal character of the kind of enlightenment given by the gift of understanding, in contrast to truths grasped by the insights of an autonomous intellect. Furthermore, the last example also suggests how Aquinas thinks this understanding might be communicated. I suggest that one would 'open' a text by *pointing out* what is important within the lesser details and making relevant connections.
38. *ST* II-II.45.3, "Sapientia quae est donum est excellentior quam sapientia quae est virtus intellectualis, utpote magis de propinquo Deum attingens, per quandam scilicet unionem animae ad ipsum."
39. Super Io., 4, 2 "Nam ipse spiritus sanctus est fons indeficiens, a quo omnia dona gratiarum effluunt; I Cor. XIII, 11: *haec omnia operatur unus atque idem spiritus* et cetera. Et inde est quod si aliquis donum spiritus sancti habeat, et non spiritum, aqua non continuatur suo principio, et ideo est mortua, et non viva."
40. See Peter Hobson's "What puts Jointness into Joint Attention?" in Eilan et al. (2005, 185).
41. A consistent failure to point out objects to others, pointing being a form of (or invitation to) joint attention, is now recognized as a classic symptom of autism.
42. Hobson in Eilan et al. (2005, 191).
43. An autistic child is, therefore, able to respond to another person in Stephen Darwall's sense of the 'second-person standpoint', that is, "the perspective that you and I take up when we make and acknowledge claims on one another's conduct and will." See Darwall (2006). For an introduction to the main ideas and examples from everyday life, see pp. 3–38. Douglas Lavin has provided a helpful review and critique of Darwall's thesis for the *Notre Dame Philosophical Review*, January 2008.
44. Such findings are consistent with other symptoms of autism in young children, such as a failure to follow the gaze of an adult toward an object, failing, in other words, to appropriate and track a second-person's stance.
45. Johannes Roessler, "Joint Attention and the Problem of Other Minds," in Eilan et al. (2005, 247).
46. For a comparison of pronoun reversal in autistic and Down syndrome children, see, for example, Tager-Flusberg (1993, 184).
47. Based on a recent communication with Prof. Peter Hobson, I have adopted the term 'relatedness' here, reserving the term 'relationship' to a possible fruition

of such relatedness. In the draft of a project proposal we are working on, Hobson pointed out, "The word 'relationship' conjures up a form of meaningful connectedness that is established over time with a particular individual or individuals, perhaps a friend, an enemy, an attachment figure, or a work colleague. The idea of 'relatedness', although intertwined with that of 'human relationship', also encompasses the many and commonplace ways we relate to other persons we hardly know, and in relation to whom we could hardly be said to have a 'relationship'. The unfamiliar child towards whom I feel supportive, the person to whom I feel gratitude for giving me directions, the bus conductor towards whom I feel sympathy, are examples of such relatedness." While joint attention can involve a relationship, the term 'relatedness', as Hobson suggests in this passage, is often more appropriate.

48. Stump highlights the distinctiveness of second-person relatedness with other persons, especially its connection with narrative, in Stump (2010), chap. 4.
49. See also, for example, QDC, a.4, *ad* 11; a.8, *ad* 16 and QDM, q.5, a.5. The centrality of friendship with God as a principle of Aquinas' work highlights another distinction between the virtue ethics of Aquinas and that of Aristotle. Although Aristotle suggests the possibility of attaining a kind of similitude of divine activity, he denies that a human being can be friends with any god. See *Nicomachean Ethics*, X, 8, 1159a3-9.
50. *ST* I-II.11.3: "Ad rationem fructus duo pertinent, scilicet quod sit ultimum; et quod appetitum quietet quadam dulcedine vel delectatione … Quod ergo est simpliciter ultimum, in quo aliquid delectatur sicut in ultimo fine, hoc proprie dicitur fructus, et eo proprie dicitur aliquis frui."
51. Aquinas argues, for example, that certain specific fruits, such as joy and peace, are *operationes* and not virtues (*ST* II-II.28.4; II-II.29.4).
52. In Gal 5.6. Regarding terminology, Aquinas uses the word *actus*, for example, when he argues that joy and peace are *actus* of *caritas*, and when he describes a fruit as an *actus* of virtue rather than a virtue (*ST* II-II.28.4). Aquinas uses the word *operatio*, for example, when he argues that counsel has no assigned fruit (*ST* II-II.52.4 *ad* 3) and when he describes a fruit as a kind of perfect *operatio* (In Gal, 5.6). Sometimes Aquinas will use the term *opus*, when, for example, he describes the fruits as *opera* of the virtues and of the Spirit (*opera virtutum et spiritus*) (In Gal, 5.6). As the usual translations (such as 'act', 'operation' or 'work') can be misleading when discussing the fruits, I have not translated these words, especially as the determination of the nature of the fruits is one of the goals of this paper.
53. *ST* II-II.29.3: "Duplex unio est de ratione pacis … quarum una est secundum ordinationem propriorum appetituum in unum; alia vero est secundum unionem appetitus proprii cum appetitu alterius. Et utramque unionem efficit caritas …Et propter hoc inter amicabilia unum ponitur identitas electionis, ut patet in IX Ethic.; et Tullius dicit, in libro de amicitia, quod *amicorum est idem velle et nolle*."
54. In *ST* II-II.29.3.Aquinas is principally referring to the alignment involved in the friendship of two human persons, but *caritas* also signifies friendship with God (I-II.65.5), implying that the peace involved in the love of God is also a kind of alignment.

55. In *ST* I-II.52.4, Aquinas assigns no fruits to the gift of counsel, even though counsel helps direct or bring about the actions of other fruits, such as goodness and benignity. Counsel is, therefore, an instance of a gift for which the corresponding *actus* are not fruits.
56. In the case of peace, for example, the beatitude of peacemaking precedes the fruit of peace (*ST* II-II.45.6).
57. Dante Alighieri, The Divine Comedy, trans. Henry Wadsworth Longfellow (Boston 1867).
58. When a playground swing, for example, is given a periodic push at a certain frequency, the natural frequency of the swing, it is very easy to make the swing go higher. In the terminology of the physical sciences, the two systems, the swing and the person pushing, are said to be in resonance.
59. The notion of resonance and *habitus* matching is also hinted at by Aquinas' description of the fruit of Benignity. Aquinas twice explains that 'Benignity' means 'good fire' (*bonus ignis*), one by which a person 'melts' to relieve the needs of others (*ST* I-II.70.3; In Gal 5.6). Given that a 'good fire' is also one of the most common symbols of the Holy Spirit, Aquinas seems to imply that the person becomes *like* God in the manner in which she loves others.
60. Aquinas examines prayer as an act of the virtue of religion annexed to the virtue of justice, *ST* II-II.83. Although humility might not, at first, seem to pertain to friendship, the various species of pride, to which humility is opposed, turn out in practice to involve some kind of impairment of friendship. A person given to empty boasting, for example, is not disposed to friendship, and someone who considers that he has made himself great, or to be deserving of greatness, will not be amenable to friendship with a personal agent who has bestowed such attributes as free gifts. For the species of pride, see *ST* II-II.162.4.
61. Porter makes precisely this point, "The infused virtues function in a way that is significantly different from the way in which the acquired virtues function, so much so that they can be described as virtues only in a carefully qualified sense" (Porter 1992, 20).
62. The notion of 'reconciliation' in this example has obvious parallels with one of the alternative names for the Sacrament of Confession, one of the effects of which is held to be a restoration of the infused virtues and gifts.
63. See, for example, Hobson et al. (2006).
64. As Stephen Angle has argued, character and dispositions, rather than rules, are foundational to human flourishing in both classical Confucianism and Neo-Confucianism. See Angle (2009, 53–55).
65. Angle (2009, 61–74). Angle also notes that Confucian thought draws attention to the role of empathy in a good life and there is a suggestion that, when individual sages attend together to the 'coherence' of things, these sages will also have a kind of unity, even without explicit communication (Angle 2009, 45).
66. The ideas presented in this chapter are explored in more detail in a recently published book, *The Second-Person Perspective in Aquinas's Ethics: Virtues and Gifts* (Pinsent 2012). I would like to thank Eleonore Stump for introducing me to the potential of second-person relatedness for understanding ethical problems, and for reviewing certain ideas of this chapter presented in a brief

chapter "Gifts and Fruits," recently published in *The Oxford Handbook to Thomas Aquinas*, edited by Brian Davies and Eleonore Stump (Davies and Stump 2012). I would also like to thank the organizers of the Logos Workshop at Rutgers University, 13–15 May 2010, sponsored by the Center for Philosophy of Religion at Notre Dame and the Department of Philosophy at Rutgers University, for the opportunity to receive feedback from a preliminary presentation of these ideas. I also acknowledge, with gratitude, that this work was made possible through the support of a grant from the John Templeton Foundation. The opinions expressed in this publication are those of the author and do not necessarily reflect the views of the John Templeton Foundation.

References

Alighieri, Dante. 1867. *The divine comedy*, 3 vols. Trans. Henry Wadsworth Longfellow. Boston: Ticknor & Fields.

Angle, Stephen C. 2009. *Sagehood: The contemporary significance of neo-Confucian philosophy*. New York/Oxford: Oxford University Press.

Aquinas, Thomas. 1964. Natural and supernatural. In *Summa Theologiae: Volume 1, Christian theology: 1a. 1*, ed. Thomas Gilby, 99–101 (Appendix 8). London/New York: Eyre & Spottiswoode/McGraw-Hill Book Company.

Aquinas, Thomas. 1981. *The Summa Theologica of St. Thomas Aquinas*. Trans. Fathers of the English Dominican Province. Notre Dame, Ind.: Christian Classics. First published 1911–1935. London: Burns, Oates and Washbourne Ltd.

Darwall, Stephen. 2006. *The second-person standpoint: Morality, respect, and accountability*. Cambridge, MA: Harvard University Press.

Davies, Brian, and Eleonore Stump (ed.). 2012. *The Oxford Handbook of Aquinas*. New York: Oxford University Press.

DeYoung, Rebecca Konyndyk. 2003. Power made perfect in weakness: Aquinas' transformation of the virtue of courage. *Medieval Philosophy and Theology* 11(2): 147–180.

Eilan, Naomi, Christoph Hoerl, Teresa McCormack, and Johannes Roessler. 2005. *Joint attention: Communication and other minds: Issues in philosophy and psychology*. Oxford: Oxford University Press.

Flüeler, Christoph. 2007. Teaching ethics at the University of Vienna: The making of a commentary at the Faculty of Arts (A case study). In *Virtue ethics in the middle ages: Commentaries on Aristotle's Nicomachean ethics, 1200–1500*, Brill's studies in intellectual history, ed. Istvan Pieter Bejczy. Leiden: Koninklijke Brill NV.

Garrigou-Lagrange, Réginald. 1946. *La Synthèse Thomiste*, Bibliothèque française de philosophie. Paris: Desclée de Brouwer.

Hause, Jeffrey. 2007. Aquinas on the function of moral virtue. *American Catholic Philosophical Quarterly* 81(1): 1–20.

Hobson, Peter, Gayathri Chidambi, Anthony Lee, and Jessica Meyer. 2006. *Foundations for self-awareness: An exploration through autism*, Monographs of the society for research in child development 284, vol. 71. Oxford: Wiley-Blackwell.

Inglis, John. 1999. Aquinas' replication of the acquired moral virtues. *Journal of Religious Ethics* 27(1): 3–27.

John of St. Thomas. 1951. *The gifts of the Holy Ghost*. Translated by Dominic Hughes O.P. New York: Sheed & Ward.

Jordan, Mark. 1993. Theology and philosophy. In *The Cambridge companion to Aquinas*, ed. Norman Kretzmann and Eleonore Stump, 232–251. Cambridge: Cambridge University Press.

Kanner, Leo. 1968. Autistic disturbances of affective contact. *Acta Paedopsychiatrica* 35(4): 100–136.

Kent, Bonnie. 1995. Does virtue make it easy to be good? The problematic case of St. Paul. In *Les Philosophies Morales et Politiques au Moyen Âge: Actes du IXe Congrès International de Philosophie Médiévale, Ottawa, 17–22 Août 1992*, vol. II, ed. Bernardo C. Bazán, Eduardo Andújar, and Leonard G. Sbrocchi, 723–732. Ottawa: Legas.

Kent, Bonnie. 2002. Scotus on virtues. In *The Cambridge companion to Duns Scotus*, ed. Thomas Williams. Cambridge: Cambridge University Press.

MacIntyre, Alasdair. 2007. *After virtue: A study in moral theory*, 3rd ed. Notre Dame: University of Notre Dame Press.

Miner, Robert C. 2000. Non-Aristotelian prudence in the Prima Secundae. *The Thomist* 64(3): 401–422.

Miner, Robert. 2009. *Thomas Aquinas on the passions: A study of Summa Theologiae, 1a2ae qq. 22–48*, 1st ed. Cambridge: Cambridge University Press.

Park, Clara Claiborne. 1967. *The siege: The first eight years of an autistic child.* New York: Little, Brown and Company.

Pinckaers, Servais. 2003. *Morality: The Catholic view*. South Bend: St. Augustine's Press.

Pinsent. 2012. *The second-person perspective in Aquinas's ethics: Virtues and gifts*. New York: Routledge.

Porter, Jean. 1992. The subversion of virtue: Acquired and infused virtues in the "Summa theologiae". *Annual of the Society of Christian Ethics* (1992): 19–41.

Shanley, Brian J. 1998. Divine causation and human freedom in Aquinas. *American Catholic Philosophical Quarterly* 72(1): 99–122.

Stump, Eleonore. 1999. Wisdom: Will, belief, and moral goodness. In *Aquinas' moral theory: Essays in honor of Norman Kretzmann*, ed. Scott Charles MacDonald and Eleonore Stump, 28–62. Ithaca: Cornell University Press.

Stump, Eleonore. 2010. *Wandering in darkness: Narrative and the problem of suffering*. Oxford: Oxford University Press.

Tager-Flusberg, Helen (ed.). 1993. *Constraints on language acquisition: Studies of atypical children*. Hillsdale: Lawrence Erlbaum.

Tanquerey, A. 1930. *The spiritual life: A treatise on ascetical and mystical theology*. Tournai: Desclée Company.

Aquinas on Shame: A Contemporary Interchange

Thomas Ryan

We all blush, feel uncomfortable, and for a host of reasons. As part of life, shame "raises questions of great and enduring interest concerning what it means to be human." These words encapsulate Elspeth Probyn's concern in *Blush: Faces of Shame* (Probyn 2005, xviii).[1] This is an enlightening, enjoyable and even uplifting work, due to its accessible scholarship, personal engagement and, at times, the author's courageous transparency.

In accepting the author's invitation to come exploring with her, I found myself following her suggestion to discover 'sidetracks' of my own. At times, I kept coming back to shame as a moral emotion. Understood from this perspective, its natural place is in an ethics of virtue with the focus on the moral agent's character rather than one that highlights action—as does an ethics of duty. Yet, there is divided opinion about shame's role in the moral life. Cheshire Calhoun observes that some moral philosophers consider that it is a "more primitive and less useful moral emotion than guilt" and that individuals and cultures should move past it. Shame as a moral emotion and the public exposure involved seems less directed at any wrong done than at how we appear or "at what *other people* require us to do or like" (Calhoun 2004, 127–28).[2] In other words, I am dependent on others for my sense of moral worth and what I should do to achieve that. Calhoun takes the opposite position, arguing that, if one is a participant with others in a life of shared moral practices, then shame over moral failings is "essential to a mature moral agent's psychology" (Calhoun 2004, 129). This approaches the position of Thomas Aquinas on shame and its role in the moral life. In their investigations, then, both Calhoun and Probyn resonate with the work of Thomas Aquinas on shame.

My approach will use three key positions argued in *Blush* (Probyn 2005, ix–xviii) as markers to re-visit Aquinas' discussion of shame in his *Summa Theologiae* (henceforth *ST*).[3] According to Probyn, shame: (a) consistently entails values,

T. Ryan (✉)
Faculty of Theology and Philosophy, Australian Catholic University,
Banyo, Brisbane, Australia
email: tryansm@bigpond.net.au

T.B. Mooney and M. Nowacki (eds.), *Aquinas, Education and the East*, Sophia Studies in Cross-cultural Philosophy of Traditions and Cultures 4,
DOI 10.1007/978-94-007-5261-0_5,

self-evaluation and what it means to live a good life; (b) is an integral part of healthy human functioning in the personal and social realms; (c) can be seen as a universal, even 'essential', aspect of human life. I will compare and contrast Probyn and Aquinas in three stages: first, in relation to their respective contexts, aims and methodologies; second, in their understandings of the meaning and role of shame; thirdly, in identifying some subversive aspects underlying Aquinas' treatment of shame and some contemporary implications across cultures. I conclude with some brief observations about the role of shame in contemporary moral education.

1 Contexts, Aims and Methodologies

Clearly, Probyn and Aquinas have differing historical contexts. The backdrop to Probyn's book is the world shaped by rapid change, post-modernism, multiple perspectives, pluralism of cultures, and especially that of feminist thought. Further, her methodology blends the empirical (quantitative research) and the qualitative (personal experience and narrative) with insights from sociology, psychological theory and cultural anthropology. Her conclusions and arguments have their grounding in researched data and informed commentary.

Aquinas, too, lived in a period of economic and social change. He was a member of a new religious order (Dominicans) founded to meet needs centered on an emerging urban economy and increasing mobility. He was open to other cultural perspectives (e.g., the Islamic). Yet, in contrast with the 'historical consciousness' found in Probyn, the world-view of Aquinas generally manifests the stability and order characteristic of 'classical consciousness.'[4] His primary focus as a teacher was the exposition of the Hebrew and Christian Scriptures (*Sacra Pagina*). In his major scholarly work, the *Summa Theologiae*, his primary aim is to elaborate a theological synthesis of the Christian faith within an ecclesial context. In this task, he gives a special place to the tools of philosophy, especially of metaphysics and philosophical psychology. His writing appears, at times, to be informed by his personal experience, as in his insightful calibrations of love and friendship.[5] But, overall, his work is characterized more by philosophical argument from reflection on common experience than by rigorous empirical method or the warmth of personal narrative.

Again, while Probyn acknowledges the role of other emotions and affective realities, her dominant focus is on shame and its role in human life. By contrast, Aquinas has a broader canvas, namely that of virtue and the Christian life.[6] Within that framework—the equivalent of two or three books—he develops, on the theological foundation of divine grace, a moral psychology of the emotions, the affective virtues and their integral role in moral action and Christian discipleship.

Between the two authors, there is an interesting point of convergence. Probyn presents two approaches to shame. As a psychological/scientific reality, shame is an 'affect' that involves the workings of the brain and associated bodily reverberations. As a sociological/cultural reality, shame is an 'emotion' which has a cognitive component and is expressed socially.[7] Interestingly, this approach has its parallel in

Aquinas' recognition of the psychosomatic aspects of human behaviour. He uses 'passions' (*passiones*) to describe movements of what he names the 'sensory appetite' (*appetitus sensitivus*)—the bodily aspect of human affectivity that is 'affected' or 'moved' to be immediately responsive to sense experience, particularly in the area of relationships. For Aquinas, the three key elements of this process entail (a) an apprehension or belief about an object; (b) an evaluative cognition of an object, namely, an attitude for or against an object perceived to be agreeable/disagreeable, good/ bad (in some way) and (c) a bodily alteration associated with, but not identified with, the emotional response.[8] The emotional response is a movement either *towards* the object, as in love (*amor*) or pleasure (*delectatio*), or *away* from the object as in hate (*odium*) or fear (*timor*).[9] For Aquinas, 'passion' is a blend of 'affect' and 'emotion' found in Probyn.

2 Meaning and Role of Shame

Consider four moments when we find ourselves embarrassed. I walk into a room and think someone else is smiling at me. My interest is aroused. I move forward to talk to the person and realize that the smile was directed to someone just to my left. I had misread a cue. My recognition and interest were misplaced. I feel awkward and self-conscious.

Again, I walk through a half-closed door. I find two people in a hushed conversation. I instinctively say 'sorry' and withdraw. Or it may be that the same couple are embracing or even engaged in sexual intimacy. My discomfort is more intense. I quickly stammer a blushing apology then make a fast retreat from a similarly red-faced couple.

Thirdly, I see a public figure covering his face with a newspaper on television. I am not sure if he is a convicted criminal leaving a courthouse or an innocent citizen hounded by paparazzi. Whatever the case, I see an instinctive urge to hide from public gaze.

The final example involves a change from imagined scenarios to an actual experience of childhood told by Probyn. She recalls how, as an 8 year old, she made another girl cry. She did this by teasing her because she did not have the same name as her mother. The author recognizes that she was a child. She could not have appreciated that the little girl's mother had remarried and taken another man's name. But, even years later, this does not stop Elspeth Probyn from blushing. As with the other fictional examples above, we can readily identify with the sense of feeling small and undone. Perhaps, too, we can identify with Probyn in that her story may prompt a flash of memory of a shameful moment from our own childhood.

What do these scenarios have in common? Probyn suggests that what they have in common (blushing) is "the body calling out its interest" (Probyn 2005, 28) which manifests the desire for *connection*, "to fit in" (Probyn 2005, 38).[10] In these examples above, there is a sense that it is my body telling me that *I am out of place*. In the first example, I have not been invited to join the person who seems to smile in my direction. In the second example, in entering the physical space of two other people, they feel invaded and I, an intruder. In the third scenario, beneath the desire to hide from public exposure (even if they are innocent) lies the desire not to be excluded but to belong. In the final example, by being demeaned and humiliated, the child was made to feel 'on the outer', as someone who did not fit in. Probyn, on the other hand, felt diminished in her diminishing of the other person. She was 'out of place', in an analogical sense, in that she was not the person she would like to be. There is also implied that, with an adult's perspective, Probyn had some sense of being able to put herself in the other child's 'place', to somehow identify with what the other little girl felt at the time.

Common to these examples are the three 'moments' noted above from Aquinas' treatment of an emotion: a cognitive aspect, i.e., an apprehension or belief; an evaluative 'judgment'; finally, a 'being affected' in a bodily and psychological way. From an awareness of a situation, I am *emotionally* moved through my body (I blush) in sensing that a boundary has been transgressed within the realm of those social/cultural or moral patterns of how to act or not to act. I feel exposed. I want to hide, even from myself and my mistakes or my deficiencies. This brings us then to our first marker.

Shame consistently entails valuation, self-evaluation and a framework for how to live a good life.

We have noted above that, common to the four scenarios, is that blushing is "the body calling out its interest" (Probyn 2005, 28), and, as Probyn suggests, that the various 'faces of shame' reveal our desire for *connection* and even the possibility of love.[11] When connection or love appear to be offered but are not, one feels exposed and even rebuffed. The same is true when one shows interest in another and it is not reciprocated. Again, we all know that 'mortified' feeling if others know something we have done. That feeling is even more intense if we are caught in the act. What others think of me and how I think of others is important. Shame has a relational and communal context. Shame is a register of those connections and the interests they involve. One instinctively recognizes that there is something very wrong with a person who feels no shame.[12]

Probyn argues that shame and its accompanying interest (in myself, in others) entails what is *important* to me—the things and people I hold dear. How is it related to guilt? The guilt that is triggered by specific acts will often be accompanied by shame. Guilt acknowledged can be eased by an apology, perhaps forgiveness, sometimes reparation. Probyn suggests that both guilt and shame can be "excited by what others think about us" but that "shame goes further…to how we think about ourselves" and often demands a global "re-evaluation of the self" (Probyn 2005, 45). In that sense, she argues, shame "lingers deep within the self" (Probyn 2005, 2, 45–6). One may not feel guilty about a past action because reparation has occurred and a relationship has been restored. Nevertheless, as Probyn notes, shame about the action can "revisit us long after the particular moment of shaming has passed" (Probyn 2005, 46).[13]

So when I am ashamed it is because of my strong interest to be a good person, says Probyn.[14] Shame brings a temporary feeling of being "more fragile in ourselves" and that acknowledging this "may serve as a basis from which to reevaluate one's existence" and hence "entails a self-transformation" (Probyn 2005, 64).[15] At the same time, it is revelatory—disclosing our "values, hopes and aspirations, beyond the generalities of good manners and cultural norms" (Probyn 2005, x). It may even imply a radical shift in attitude and in embedded patterns of responding and acting ("rerouting the dynamics of knowing and ignorance") (Probyn 2005, 105). There is a bridge between personal life and cultural practices. Thus shame comes, in a sociological term from Pierre Bourdieu, within the domain of *habitus* which he defines as a non-discursive knowing or "embodied history, internalized as second nature and so forgotten as history—the active presence of the whole past of which it is the product" in which the body is "a repository for the social and cultural rules that, consciously or not, we take on" (Probyn 2005, 51, xvi).[16]

Through shame, according to Probyn, we are consistently reminded that we are embodied beings. Whether at the interpersonal, social or cultural level, shame points to boundaries, to habitual patterns of how we see values and rules and respond to them. These boundaries and patterns are reflected across cultures in the experience of the body 'being out of place.' For instance, Probyn tells of a journey to Central Australia and a visit to Uluru. She feels caught by a sense of being an outsider, 'I don't belong here.' She was intruding on another's space, that of the indigenous inhabitants of the continent.

Again, this sense of 'being out of place' is evident in awareness of what is private, even sacred, for instance, in the sexual area. Probyn says that sexuality (sexual identity) is commonly held "as an area ripe for shame." But she notes that it is not necessarily a "site of shame" or "the same site of shame for everyone" (Probyn 2005, x). Later, she writes of the people of Mt Hagen who speak of big *pipil*—the shame accompanying sexual activity in public or incest (Probyn 2005, 32). While Probyn does not investigate sexual activity in depth, we are reminded that there are boundaries and norms of acceptable behaviour concerning the exercise of sexual activity in every culture. Its accompanying sensitivities are trampled over by, for instance, pornographers who "parade the vital privacies" (Ward 2002, 159)[17] and expose the whispered vulnerabilities of sexual experience to public gaze.

Finally, Probyn rightly stresses that shame makes us reflect on who we are and what our actions might set in motion (Probyn 2005, 8, 34). We have noted how acknowledging the feeling of 'fragility' accompanying shame may trigger a re-evaluation of one's existence. Hence, shame's positive role as self-evaluative and self-transforming emerges *only if it is somehow acknowledged*.[18] As with any 'negative' emotion, there is the need for conscious engagement with shame if it is to contribute to human well-being. Shame is integral to self-assessment. To live a good life involves deliberation and freedom. This is particularly the case if shame is not to remain one expression of Bourdieu's understanding of *habitus* in which internalized cultural rules hold individuals in a pattern of immature or self-destructive behaviour. For shame to move into the realm of adult conscience and a sense of personal responsibility, a basic level of self-awareness is required. Clearly, for Probyn, shame as a cultural and psychological reality is one of life's teachers—in the social, relational and moral domains. This approaches the position of Aquinas as he builds on Aristotle's ethics by approaching shame in the setting of virtue and the development of 'character' through chosen self-direction in one's life according to the requirements of authentic moral wisdom.[19] Aquinas' primary context is shame as integral to the moral life, and this will, inevitably, have cultural and social implications.

For Aquinas, emotions are essential to the moral life and human integration. Shame, closely associated with the body (especially touch), is part of the affective virtue of temperance or self-care concerning our bodily, sexual and affective needs.[20] Aquinas' approach, contra the Stoics, is that of an average sensual person for whom friendship with God entails enjoyment and harmony in mental, bodily, sexual, emotional and social existence. While his method differs from that of Probyn, there are central insights that are common.

For our purposes, Aquinas' treatment of shame (*verecundia*) in the *Summa* is encapsulated at three points: in one specific question on the morality of the emotions, as a species of fear, and as part of the virtue of temperance (*ST* I-II.24.4; I-II.41.4; II-II.141–144). The morality of our emotions will be our starting point. He begins by asking (in carefully worded language) whether there is any emotion that is always good or evil "by its very nature" (*ST* I-II.24.4).[21] From an earlier discussion (*ST* I-II.24.1), he argues that any emotion's moral status is discerned, firstly, in so far as it guided by reason and, secondly, in a *relational* context.[22] In the language of traditional moral theology, an emotion, like actions generally, can only be evaluated morally in terms of its object, end and circumstances.[23]

Aquinas replies that an emotion that is good of its very nature is *shame* (*verecundia*) which is alternatively named as modesty (*timor turpis*). Both these terms denote fear of doing what is morally base in one's own eyes especially because it is damaging to oneself in the eyes of others. Citing Aristotle (1976),[24] Aquinas says that *verecundia* is a praiseworthy emotion. He notes elsewhere that it is a virtue in the broad sense (*ST* II-II.144.1) since 'feelings of shame' foster a disposition to avoid what brings disgrace or opprobrium (*ST* II-II.144.2).[25]

Timor turpis may be translated as 'modesty, or fear of unchastity.'[26] At first glance, this seems to restrict the scope of shame to restraint of sexual desire. However, As Eleonore Stump points out, for Aquinas, *castitas* (chastity) can have a broader meaning, namely, it denotes self-discipline concerning morally unacceptable desires (see *ST* I-II.70.3). She offers the example of *castitas* as restraining oneself from kicking the dog at the end of an exasperating day since the desire to take out one's frustrations by kicking the dog is a desire that is never acceptable to act upon (Stump 2005, 556, n.60). The word *castitas*, then, captures the exercise of the virtue of temperance (as will be explored later). This can entail sensitivity to anything that would undermine the harmony that temperance has as its goal and the self-respect that underlies it. More broadly, Aquinas recognizes that shame's sensitivity can concern any action or vice that would undermine or oppose one's moral excellence and bring one disgrace in the eyes of others (*ST* II-II.144.1 *corp.* & *ad* 2; and *ST* II-II.144.2). We can examine Aquinas' approach first in general and then in specific terms.

Firstly, for Aquinas, shame (*verecundia*) as an emotion is good or evil of its very nature in a relational context, namely, as being 'in tune' (*conveniens,* fitting) or 'out of tune' (*dissonans*, not fitting) with right reason and authentic humanity. Its measure is in response to the question: How would the practically wise or virtuous person respond in this situation? Shame is an emotion that enhances human flourishing, personally and socially. While shame is negative in the sense that it makes us feel uncomfortable, its positive function emerges from its object, namely the value it is directed towards upholding and the attitude produced.[27] In this case, it is moral excellence and fear of disgrace in acting against such excellence.[28] Shame, then, protects us in our deepest convictions. In contemporary terms, by disposing our sensitivity to what can distort our moral horizon, shame is a sentinel guarding our personal self-transcendence in terms of the search for meaning, truth and value.

Aquinas is arguing that certain emotions such as shame, when understood and described carefully, have a built-in significance. It is not that they are morally neutral (psychological facts) and one's attitude to them makes them morally good. It is rather that the emotion itself crystallizes an habitual disposition to make, with ease and consistency, a 'felt evaluation' of an intentional object in that it is perceived as 'fitting' (good) or 'not fitting' (evil). This is precisely the understanding of Martha Nussbaum.[29] Aquinas also holds that concerning emotions and their expression in the affective virtues (e.g., temperance, fortitude), what constitutes moderation (the mean) differs from person to person.[30] While avoiding any suggestion of emotions as responses of 'perfectly programmed' automatons (an understandable concern of Probyn's (2005, 10)), there are some 'objects' and situations that are arguably 'fitting' (right) and appropriate in our personal and social life. For instance, for both Probyn and Aquinas, to feel no shame (to be morally 'shameless') or to lack sensitivity to another's pain, is neither desirable nor admirable. For both authors, shame reveals both the values and the moral configuration of a person. Shame is an emotion that reverberates in both the intra-personal and inter-personal domains.

We return to the specific aspect noted earlier, namely to shame in relation to sexuality. Aquinas' cryptic, even elliptical, treatment assumes the reader's awareness of the broader context of his discussion. The translation of *timor turpis* as 'fear of unchastity' captures a usage of *castitas* in terms of self-control and moderation in the realm of the bodily and emotional life.[31] The integration and appropriate expression of one's sexuality is representative of such a virtue. First, Gilby notes that, for Aquinas, shame has a range of different bodily expressions (Gilby 1968, 55).[32] Second, Aquinas' treatment of the gift of sexuality is earthy and basic. He does not give any failure in the sexual area of life "the dreary eminence it has for later moralists" (Gilby 1968, xxiii). Third, he acknowledges human ambivalence in the sexual domain in terms of a certain 'powerlessness' over our emotions or sexual movements (even in the virtuous exercise of one's sexuality, see *ST* II-II.151.4). Fourth, and most importantly, it is not by chance that Aquinas' language about shame, especially in relation to the sphere of the body and desire, suggests a concern with self-respect and the sacredness of the person. The word *verecundia* (shame, modesty) has its verb root in *vereor* (respect, fear, reverence). This is foundational for Aquinas. For him, shame's object is not the body or one's sexuality but the 'out of place' (*dissonans*) invasion of any area of embodied personhood that warrants respect. Shame's scope is moral integrity and how that can be sullied and the person disgraced in some way.

Hence, only with great difficulty can we construe Aquinas' view of shame simply as fear of sexual sin (the misuse of one's sexuality). In its broader setting, shame is prompted by a sense of *respect for the self* and sensitivity to one's moral ideals and character.[33] This point becomes evident through Aquinas' later discussion in three points: (a) healthy self-love is an essential component of Christian living (*ST* II-II.25.4); (b) we must have love for our body as a gift from God (*ST* II-II.25.5); and (c) concern for one's own good is integral to virtue or moral self-transcendence (*ST* II-II.26.6). It is not surprising that, for Aquinas, while growth in virtue will mean greater attraction to the good and less reliance on shame to deter us from evil, there is still some truth in saying that the more virtuous a person is, the more they are disposed to a sense of shame about any personal failure in moral excellence (*turpis*) (*ST* II-II.144.4).[34] In other words, shame is the guardian of values that have been personally interiorized.

Probyn (briefly) and Aquinas (in his more elaborated treatment) mirror what is common to all cultures, namely, a sacred 'space' around a person as embodied and especially as a sexual being capable of love and intimacy. Shame implies reverence for vulnerability and the intimate whispers 'spoken in the night' alluded to earlier.[35] Aquinas himself speaks of a "certain delicacy" needed in sexual matters and of "a respect which is the opposite of shamelessness [which] sets up a certain reticence and sense of impropriety about exposure." (*ST* II-II.154.9) The intersection of personal and social life entails respect, care for oneself, and boundaries. Transgression of boundaries evokes an instinctual movement of shame and accompanying self-evaluation. This brings us to the second marker.

Shame is integral to healthy human functioning both personal and social.

We have noted that Probyn explores the productive role of shame "as an essential part of yourself" (Probyn 2005, x) and as something that we do 'well' together. Further, she probes the intimate connection between shame and interest. In these tasks, Probyn is indebted to the work of the American psychologist Silvan Tomkins. Shame, like fear or anger, is an emotion whose role, normally speaking, is to make us feel uncomfortable.[36] There are some things we should be ashamed of, just as there are things about which we should be angry or afraid. Like any emotion, especially those that we call 'negative', shame can be either constructive or destructive. Feeling shame can sustain personal well-being and guide our responses in our relationships and our social life. The educative function of shame can work at the collective level. In this context, used properly, shame can be a positive instrument for healing and reconciliation, as in processes of restorative justice.[37] Again, at the collective level, it has been noted earlier that there is a range of cultures where shame is an instrument of socialization and enculturation.[38] Shame's socializing function has specific implications for the person's sense of self in relation to shame and socially acceptable behaviour.[39]

On the other hand, we cannot overlook shame's capacity to undermine the sense of self. It can be an instrument of reproach, power, and submission. As Probyn notes, for already damaged individuals, shame can be "lethal" (Probyn 2005, 92).[40] Further, shame can be a powerful tool to create injustice, especially in regard to women and ethnic groups. Such groups are representative of the individual and collective historical experience of shame involving subordination as a "pervasive affective attunement to the social environment" (Probyn 2005, 83).[41]

For Aquinas, shame, as part of the virtue of temperance or self-care, helps us to grow in the likeness of God. It is reflected in sensitivity to whatever demeans oneself as a person. Its companion is *honestas*, namely, a sense of moral excellence and of love for its beauty.[42] Shame makes one more sensitive to what threatens virtue, personal goodness, and, most importantly, what fosters or undermines our responsiveness in relationships (*ST* II-II.142.4 and 144.1). Like Probyn, shame for Aquinas reflects interest in being a good person. Within the Aristotelian tradition of the virtues, Aquinas highlights the personal sphere, namely patterns of habitual response and action that are virtues or good moral habits. Complementing this analysis, Probyn's access to the social sciences enables her to analyze the dynamics of social and cultural influences on personal life, especially in their distorted and destructive forms.[43] She addresses what, in theological terms, could be understood as structural or social sin.

Aquinas certainly sees shame in terms of disapproval or loss of face with others (*ST* II-II.144.3), and, in that sense, shame is socially and culturally located. Unlike Probyn, Aquinas does not probe shame as a form of social control in terms of the sociological notion of *habitus*. However, in situating sensitivity to shame firmly within the virtue of temperance, Aquinas implies that shame's sensitivity is primarily personal and relational. Further, as Gilby notes, shame's close relationship to a sense of sin (guilt) makes it more "personal and agonizing than the fear of earning a bad name" (Gilby 1968, 55).[44] Again, shame reminds us of the reverberations of the primordial human sin and humanity's dissonance and destructive tendencies (*ST* II-II.163-5). While Probyn is conscious of shame's damaging potential, especially in culturally embedded attitudes and practices that shape individuals and behaviour, shame's theological dimensions are beyond the scope of her work.

It is helpful here to explore Aquinas further concerning two aspects raised by Probyn about the educative and formative function of shame. First, it has been noted that shame is both personal and relational. Further, I can be ashamed of an action or attitude in myself without holding my entire 'self' as shameful. But experience shows us that shame may not be confined to something specific but can spread throughout the whole self.[45]

There are clear hints in Aquinas about his attitude to this issue. He says in three places that shame is not so much fear of the very act of sin but of the resulting personal disgrace (*turpis*) (*ST* I-II.41.4 *ad* 2; I-II.41.4 *ad* 3; I-II.42.3 *ad* 4). Elsewhere in three places, he points to shame's potential to move from a particular aspect of a person's experience to become a more pervasive presence. First, we have noted above how Aquinas speaks of 'blushing' as the reaction of the whole person at the prospect of *any* behaviour or action that would be morally "disgraceful." (*ST* II-II.144.2) Again, in *ST* II-II.144.4 *ad* 3, he notes that the virtuous person (who is unlikely to do, or will readily avoid, base actions and, hence, disgrace) can be strong enough to give no weight to false reports that could lead to possible disgrace in the eyes of others. Nevertheless, Aquinas realistically points out that the same virtuous person, in maintaining his position rationally in the face of the false reports, is not impervious to being swayed by shame.[46] A stronger example is found in a previous article where Aquinas, in principle, limits the scope of moral shame to the blame and loss of face for culpable actions. However, in practice, feelings of shame and disgrace can start to envelop other aspects of one's person through the attitude of others, for instance concerning economic status, birth, job etc. (*ST* II-II.144.2 *ad* 2).[47] This is a clear suggestion about the unjust face of shame in which a person's standing in the eyes of others is based on qualities that are not relevant to the person's moral goodness. Aquinas is flagging the negative impact of cultural shame.

I have argued that, for Aquinas, the paradigmatic case of shame is the ordinary person. As he notes, "the best men and the worst men lack shame…for opposite reasons." The average person, holding a middle course, "has a love of the good, but is not wholly free from evil." (*ST* II-II.144.4 *ad* 1) For Aquinas, one can recognize shame in respect to part of oneself (an attitude or an action) together with condemnation of others, but that does not necessarily mean that the self disintegrates. One can still learn and improve. The moral life is a journey of ongoing conversion. This brings us to the second aspect of shame as a teacher.

Shame's role is to help one learn from one's mistakes. For Aquinas, shame's teaching function is evident in the experience of retrospective or 'disgrace-shame'—concerning past evil actions and their effect on one's person. This can help a person to be sensitive through prospective or 'discretion-shame' to possible future actions in which shame is part of conscience's function as a moral antenna (*ST* I-II.2.41.4).[48] Sensitivity to disgrace (in one's eyes and in the eyes of others) has already been noted earlier when Aquinas says that the more virtuous a person is, the more they are disposed 'hypothetically' to a sense of shame about any personal failure in moral excellence.[49] Shame here is a measure of an increasingly sensitized conscience to what is truly good. Or, as Schneider expresses it, 'discretion-shame' is essential for human flourishing and growth in virtue (Schneider 1977, 18–19). Both Aquinas and Schneider would agree that shame can be seen as a sentinel of moral sensitivity and integrity.

However, such learning is not in isolation. For Aquinas, its context is that of the practice of the virtues exercised, importantly, within a community of shared values possessed of a common vision. Shame's correlative is honour (*honestas*) or moral excellence that is upheld and fostered by a community and its members. When there is a failure to live up to such standards of excellence by a participant in that community's life, the members have an interest in that person's behaviour since it impacts on the common good. Aquinas offers a telling insight into how this works in practice.

Honour is the acknowledgment of a person's excellence and dishonour (a dimension of shame) is recognition of a person's defects, above all when they are one's own fault. Shame can motivate someone to want to learn from three circles of relationship in a community (*ST* II-II.144.3) and the weight of testimony to the truth found there. These are the people who matter to us, who best motivate us to overcome defects and grow in virtue.[50]

The first and outer circle includes those whom we admire and those whom we would like to admire us. We would feel more ashamed if they held us in low esteem. These persons (and institutions) are the models in the community we look to, whose wisdom and goodness are embodied in sound judgment. We can look to them for greater certainty about what is truly good and virtuous. They can help me when I am mistaken, misguided or confused.

The next circle of 'witnesses' embraces friends and associates who are closer to the circumstances and variables of one's personal situation. Shame can lead us to look to friends and associates for truthful feedback because they are in a better position to understand at a personal level.[51]

Shame (fear of disgrace) also applies with regard to truthful feedback from those in the third and most intimate circle, namely those closest to us on a day to day basis. Because they know us well, they can hurt us more. We do not want to lose their respect and esteem yet we know they can constantly call us to account. In these three circles of interaction, shame prompts openness to criticism, disagreement and the motivation to learn and change.[52]

In this regard, there is a noticeable convergence between Aquinas and Calhoun, who was mentioned at the beginning of this paper. First, Calhoun points out that giving the opinions of others "weight" (hence, the power to shame), means that one takes those others seriously "as co-participants in a moral practice." Second, Calhoun goes on to argue that shame over moral failings is "essential to a mature moral agent's psychology" and that "vulnerability to feeling ashamed before those with whom one shares a moral practice, *even if one disagrees with their moral criticisms*, is often a mark of moral maturity" (Calhoun 2004, 129). For both authors, shame (like honour as the recognition of moral goodness) mediates the relationship between a community and its members. This brings us to our third marker.

There is a universal capacity to feel Shame.

We have noted that Probyn taps different disciplines in her discussion of shame. Psychology helps to tell us "something about how our bodies dictate what we feel" (Probyn 2005, xxx). Anthropology and sociology open doors on different ways of engaging social and cultural life. More specifically, Probyn draws on anthropological studies in Melanesia and correlates them with the psychological theory and studies of Tomkins.[53] Blushing as the body 'calling out its interest' has reverberations for the self in the social world, namely "[what] shame does to bodies and what bodies do to the organization of the social" (Probyn 2005, 27). The breadth and consistency of these studies together with Probyn's self-reflections and her use of the narratives of others suggest that all humans are born with a capacity for shame.

Given the ubiquity of shame as a bodily, emotional and social reality, what is to be lost, asks Probyn, by engaging with those who approach shame using other methods and vocabulary? (Probyn 2005, 25). Studies across different disciplines indicate that there is "something terribly important in shame—it is human to feel and to do it well" (Probyn 2005, 34). All humans blush. If the gagging reflex is an instinctual function to save the species from poisoning, why too not shame? Its innateness in our bodies and its organizing impact on social relations suggests that we are, by nature, social beings (Probyn 2005, 34). In suggesting the universality of shame, Probyn acknowledges that we cannot disregard cultural differences or the risk of promoting a Western model of affect.[54]

Probyn proposes that we need to be open to the evidence that shame may be 'biologically innate' and see where that leads us. This does not imply that we all blush for the same reasons, which would imply that we experience shame in the same way or that some are not more vulnerable to shame, whether culturally or temperamentally. Why should there be any necessary opposition between what is particular and what is universal? "Why should innate or universal characteristic always reduce difference?" Probyn asks (Probyn 2005, 29). "The notion of innate affects provides a way to understand both how certain phenomena are universal to humans and also how they differentiate in their causes and expressions at an individual level and within social groups" (Probyn 2005, 29).

For Aquinas, human nature, both with its biological reality and its rationality, provide the two wings of human experience as revelatory, namely the gateway to discerning the law of human nature. Moral life is built on this foundation. Nevertheless, the various expressions of shame remind us that temperamental variation is grounded in a common human nature that has, as one of its unchanging features, its openness and malleability with respect to temperament and habit. Human nature, then, as a source of morality, is subject to much variation. Aquinas acknowledges that, beyond the very general, it is difficult to arrive at moral norms that are certain and universal when faced with so much variability and contingency in human life (*ST* I-II.94.4).

There are moral philosophers and theologians who argue that the desire for happiness underpins moral theory. Emotions such as shame disclose who we are by pointing to what affects us. The more we are moved by the 'fitting' (right) objects, the more we come to flourish. This process entails an order and harmony centred on love—in respect to oneself (self-love), in ones relationship with God, and in our relatedness to others and the world in friendship and compassion. Aquinas uses the language of fittingness, of 'being in or out of tune' (*consonans/dissonans*) to describe the workings of human rationality since, in its wider setting, a human being is born to be "attuned to everything in so far as it is created in the image and likeness of God" (Kerr 2002, 31). The ethical naturalism underlying this is teleological, is progressive in the sense of being directed towards an end and is captured by the notion of 'connatural knowledge.' This notion anticipates that found in Pierre Bourdieu, within the domain of *habitus*, namely, of affective attunement or a non-discursive knowing of the social environment. Kerr concurs noting that things are "destined to a certain fulfillment, with appointed ends, modes and opportunities" (Kerr 2002, 31), and this involves the ongoing free search for truth and value. 'Right' response and action are informed by who we ought to become (the divine image) and are paradigmatically embodied in the wise person.

Aquinas' classification of emotions is built upon a common human nature which, because it is shared, allows inferences to be drawn that apply to all humans. Cultural variations and cross-cultural differences are not controlling considerations for Aquinas. For all that, if one compares Aquinas' study of the emotions, it stands up well to contemporary studies and models.[55] He is remarkably modern in his approach to negative emotions such as fear, anger, shame, sadness, loss.[56] Carlo Leget suggests that Aquinas' formal taxonomy is "open to many cultural adaptations while reserving a primary place to the concept of love" (Leget 2004, 571).

We have discussed earlier how respect, social reputation and 'loss of face' are words associated with 'shame-culture' or 'honour-shame culture.'[57] Further comment on this is appropriate here concerning Aquinas. Aquinas' treatment of shame, like Aristotle's, is the reverse side of the treatment of honour. There is a difficulty here with translation. The Latin words *honestum* or *honestas* are often rendered as honourable or honour. But in their Greek and Latin meanings they denote 'good in itself' as in *bonum honestum* (as distinct from useful or pleasurable goods). And *honestas* is best rendered as 'moral excellence.' In English 'honourable' can denote 'worthy of honour' or respect, acknowledgement and 'honour' also refers to reputation, standing or distinction in the eyes of others.[58] For Aquinas, 'honour' is given to one who possesses virtue and is 'worthy' of respect and acknowledgement.[59] He goes on to say that reputation and public respect are external to virtue and, in fact, may be extended to a person without virtue. In essence, for Aquinas, one's moral worth does not depend on one's standing in the eyes of others. Rather, how people are regarded by others ought to be a reflection of their moral excellence.

The dynamics of shame in Aquinas work on this assumption that honour is primarily due to moral excellence. The virtues are about what is good and true. They attract us by their own inner force and beauty (*ST* II-II.145.1 *ad* 1). What Aquinas is highlighting is moral beauty and its role within the communal context that underpins his conception of the moral life. If moral excellence has a radiance that inspires others and is worthy of acknowledgement in itself, then a question arises about the role of the community. Honour and shame are social responses—one to moral goodness, the other to a failure to moral goodness. They are respectively the positive and the negative poles of 'recognition.' They manifest the shared nature of the moral life.

The need for acknowledgment by others is not necessarily the same as the desire for social reputation or fear of 'losing face.' Nor is it necessarily a sign of moral immaturity or failure to interiorize moral standards. Stocker and Hegeman suggest that lack of mutual recognition of virtue or the desire for that recognition can be, citing Aristotle, 'questionable and unattractive.' They note that "wanting recognition for virtue can be part and parcel of the desire to live a responsible and responsive life with others" (Stocker and Hegeman 1996, 291).[60]

There is another aspect to the notion of a community sharing a vision of 'virtues of common pursuit' or 'shared practices.' We have to acknowledge the comparatively homogeneous society of Aquinas' (and Aristotle's) time and the more pluralistic communities of the modern Western world. As David Putnam points out, to appear honourable before one's peers today "will much more often result in standing alone before an individual or group who does not share our values." However, Putnam then goes on to note that this "does not lessen the essential communitarian foundation of character" (Putnam 1995, 286–88).

3 Aquinas: Subversive About Shame?

In the light of the universal character of shame, its role in Aquinas' moral theory and the ubiquity of 'shame cultures' historically, one could consider Aquinas as a mirror of change and development.[61] Aquinas blends many influences from the twelfth century. Specifically, he reflects the rise of the individual or of the self, a theme traced by authors such as Colin Morris and Caroline Walker Bynum.[62] There was a growing awareness, as Morris says, of a clear distinction between "my being and that of other people" (Morris 1987, 47).[63] For Aquinas, it is not the modern self characterized in terms of autonomy, namely of the self over and against the other. For him, God's image is realized in the wisdom and virtue of a loving self that is responsive in the world of relationships.

Further, Aquinas himself was shaped by his social environment as a Dominican friar. Democratic processes were present at the beginnings of the Dominican culture and experience. This was evident, for instance, in the community's role in governance (through the Chapter) and in the move away from the monastic model of leadership which centered on the Abbot. Understandably, there are traces of this democratization in Aquinas' theological method and anthropology. For instance, Aquinas holds that the human person flourishes as the image of God when intellect, will, emotions and body work collaboratively.

Our considerations prompt a question concerning the relation between collective, kin-based cultures and those that are more individually orientated.[64] Aquinas appears to straddle two worlds. His life as a Dominican friar reflects ecclesial movements adapting to the massive social and cultural shifts of the Middle Ages—from rural to urban economies, and the development of trade and exploration (both intellectual and geographical) which brought greater social mobility. Ormerod suggests that, in Aquinas, one can detect the transition from a cosmological culture in which "an individual is ordered to the society and the society to the cosmic order, the divine court" to a more anthropological culture in which "society is ordered to the individual and the individual to some world-transcending measure, such as reason" (Ormerod 2007, 232).

Aquinas, then, does not stand apart from his time and culture. For instance, Torrell points out that Aquinas remained 'feudally tied' to his milieu and his time—something reflected in his use of the "vocabulary and metaphors of chivalry and the military profession" (Torrel 1996, 12). But we can also detect in Aquinas hints of the emerging modern self. One side to this 'self' is the sense of personal responsibility and self-direction which will be explored in the interiority, for instance, of Shakespeare's Hamlet. Alternatively, the same 'self' points to two differing conceptions of morality, namely, one based on the heroic virtues centered on honour, the other on goodness, love and compassion.[65] Could Aquinas be seen as a mirror of an incipient differentiation of consciousness, of a shift in cultural self-transcendence in the representation of truth and the appreciation of value? In that process, there seems to be a subversive side to Aquinas, and that for five reasons.

First, we have discussed above Aquinas' treatment of the motivating force of shame in education in the virtuous life. This was specifically in reference to how this leads one to learn from those who are 'witnesses' to the truth. Ultimately, for Aquinas, the Christian looks first to be recognized and honoured in God's eyes, and that honour and glory come from God (*ST* I-II.3). Aquinas' moral vision is centered on and epitomized in Jesus Christ, who, as the Gospels attest, himself resisted the clean/unclean dualism of a shame society. The irony is that the symbol of the

Christian message as testimony to divine truth is Christ crucified—the starkest symbol of evil at its worst. The Christian gospel is summed up in God's love in Jesus embracing shame and ignominy at life's darkest moment—as a criminal within his social context. Shame itself is subverted and transformed, in Christ crucified, into a sign of hope. Aquinas himself acknowledges the effect of this in the attitude of the Apostles. In the Acts of the Apostles (5:41), through their magnanimity, Peter and his companions 'rejoice' to suffer public disgrace (shame) 'in the name of Jesus' (*ST* II-II.144.2 *ad* 1).

Second, the controlling benchmark of Aquinas's moral theory is neither honour, nor shame, nor even autonomy. While they have a role, the moral life is ultimately guided and animated by love. By putting the dynamics of a love that transcends shame first and foremost, Aquinas subverts Aristotle's notion of friendship and its ethical implications.[66] For Aristotle, complete friendship was a relationship between equals. Without belief in human deification in Christ, Aristotle could not envisage friendship between God and human beings.[67] Further, he could not imagine such friendship, with its roots in love for God, extending to love of enemies or forgiveness of hurt done. For Aquinas, God's power is most evident in divine mercy and compassion (see *ST* II-II. 30.4) and we are called to grow in the image of God.[68] For Aquinas, the nature and scope of virtue and its full realization through sharing in divine wisdom goes well beyond that of Aristotle.

Third, the early Church communities attempted to offer a social order that contrasted with their surrounding cultural context. Aristotle's moral community without women or slaves was superseded. Customary social ranking and status were replaced by a relationship of equality.[69] Noble, landowner, slave, freeman, woman, artisan, rich, poor—all were included. This shift is perhaps best reflected in the way that reality is portrayed in the Gospels. In Greek and Roman literature, Peter, the peasant fisherman, would be regarded as a fool, and could be treated realistically as the subject of satire or comedy, but would never be taken seriously as an heroic or tragic figure. Social expectations concerning the nature of proper shame would restrain Peter the fisherman in any way coming to the centre of the stage. Similarly, literary practice and the distinction of styles would place constraints on his treatment as a character.

However, Erich Auerbach has argued that in the New Testament, specifically in the episode of Peter's denial of Jesus as in Mark 14:66–72, for the first time in Western literature, ordinary people inhabiting an everyday world are treated with high seriousness (Auerbach 1974, 41–49). Peter is no longer an ignorant peasant without standing or interest for the reader. He is at the centre of the stage with the spotlight on him. Auerbach suggests this is a turning point in Western consciousness. Peter is the 'image' of humanity, now transformed through Jesus Christ who became incarnate into the humblest social station. His life was among ordinary people and ended with the ignominy of his Passion and Death. The power of Christ's Risen presence transforms the lives and practice of the early and subsequent Christian communities and their view of reality. Over a millennium later we find, embedded in Aquinas' moral vision, the implied, even if not realized, revolution of social and cultural arrangements reflected in the Gospels. His work is a mirror of how shame, its focus and contours, is being transformed.

Fourth, one could argue that Aquinas adumbrates the transition from shame as a form of social control to one wherein shame reflects the sensitivity required in a healthy personal and social life. Aquinas seems to anticipate the democratic sensibility within which Probyn is situated. These aspects of shame cannot be separated from Aquinas' approach to the intellectual life and to education as an ongoing discovery within a conversation. First, this process is one of growth in virtue through an interactive form of learning and practice within a community that offers 'significant others' or 'paradigmatic models.' Second, on a larger stage, his engagement with Islamic thought is representative of his openness to other philosophical and religious traditions.[70]

Fifth, Aquinas' willingness to learn from other traditions converges with what was noted earlier about cultures learning from each other (Maori and Australian indigenous communities and 'community' or 'circle' sentencing).[71] There is also an overlap with earlier comments about the dangers of a form of ethnocentricity about culture—the implicit assumptions of the superiority of western/individualist culture over eastern/collectivist forms. It takes us back to Calhoun's comment that some philosophers consider shame as a "more primitive and less useful moral emotion than guilt" and that individuals and cultures should move past it.[72] This intersection or comparison of cultures is again 'subverted' first by the Christian narrative and then by Aquinas, through a corrective balance between two extremes.

First, the bringing together of the collectivist and the individual is reflected in Jesus Christ himself in his life and teaching. He was part of a collectivist culture centered on kinship, where identity is defined by the group and where the self-worth of the person or the group is expressed in honour (Arbuckle 2010, 153). It is consistent with this context that Jesus asks his disciples "Who do people say I am?" (Mark 8:28). He balances respect for his cultural traditions yet, at times, tries to move beyond them, or rather to a higher level of integration of the social and the personal. In this, Jesus does not appeal to honour or the attitudes/acceptance of others (kin, family) as the ultimate benchmark of his moral vision. As has been noted, that benchmark is love which animates both the individual and the group and is the source of unity and growth. Jesus sees the core metaphor for this in the 'heart'—in biblical terms, the deepest level of the self where mind, will and emotions converge and direct one's choices, the locus of personal conviction.[73]

Second, an effort to find a balance between the individual and the collective is reflected in Aquinas and can be traced back to Aristotle and the Greek understanding of honour and shame. It was discussed earlier how honour and shame are social responses, the positive and negative poles of 'recognition' that manifests the shared nature of the moral life. Within that framework, Tombs, building on Bernard Williams, questions the sometimes simplistic approach to 'shame' versus 'guilt' cultures and whether there is an inevitable conflict between shame values and autonomy (Tombs 1995).[74] Tombs notes justified hesitations concerning those who view shame solely in terms of a loss of social reputation versus guilt in relation to a self defined by rational autonomy without reference to 'character'. One can feel both guilt and shame for an action, but shame looks to "what I am" and reflects a fuller understanding of "personal and social identity" (Tombs 1995, 29).[75] As Tombs notes, "by the later fifth century the Greeks were able to distinguish a shame that was governed only by public opinion from a shame that was guided by inner personal conviction" (Tombs 1995, 29). Our considerations have shown this was precisely the understanding of shame found in Aquinas building on Aristotle.

Hence, one needs to ask whether 'losing' or 'saving' face in collectivist (i.e., some eastern) cultures can be reduced simply to a desire for social reputation? Alternatively, is there a lurking assumption that an 'honour-shame' culture is unable to lead a person towards ownership of values at the level of personal conviction? It would seem that, in a collectivist culture, it is around the 'face' that respect for oneself and others constellate in so far as these are the foundation of a good life and responsive relationships. This is perhaps a richer view of personal identity than one dominated by autonomy and rationality. Do we see here another culture's expression of wanting recognition for virtue as part and parcel of the desire to live a responsible and responsive life with others? In other words, as noted earlier,[76] to 'blush' from the thought of going against one's deepest convictions (within a community of 'shared practices') is not a prerogative of any one culture.

What significance does our discussion have? Perhaps today's Western emphasis on autonomy and personal choice finds a needed double counterbalance in Aquinas' approach to shame and its grounding in the virtues and the community. First, he shares with so many world cultures (especially in the East) what can be termed a *sapiential* view of the moral life. It is the search for, education in, and practice of, wisdom. Second, friendship is his ruling paradigm—with God and others. He sees shame within a framework of relationships that are called to be equal and mutual rather than unequal and hierarchical. Yet precisely as relationships animated by love and the Spirit, they have an internal impulse to expand in scope and depth, namely, and they have an *inclusive* and *pluralistic* impulse. In other words, in Aquinas we can detect the beginnings of a shift from 'saving face' to 'facing the other.'

4 Final Observations

Probyn and Aquinas would agree that "blushing is the body calling out its interest" (Probyn 2005, 28). The body is a register of the whole person, spatially, psychologically, socially and also morally. The blushing body reveals the destructive and constructive poles of shame. On the one hand, shame involves the fear of being exposed and defenseless. This process can even be to the point where the sense of self is eroded. On the other hand, shame can intimate and protect personal dignity and goodness. Overall, there is a convergence between shame, values, and well-being both personal and social. As Gerald Coleman suggests, "Since our capacity to know what we are feeling and to experience those feelings is rooted in bodily experience, to be ambivalent about or alienated from our bodies is to be estranged from ourselves" (Coleman 1992, xv).

What emerges from the interplay between shame and culture? Probyn explores how we can overtake Aquinas through investigating how we can be out-of-place by stumbling into "other people's history, culture and beliefs of which we are ignorant"

(Probyn 2005, xvi, 94–99). Such an unexpected 'step' can be as much into Aboriginal, African or Asian expressions of a collectivist culture. Such an experience of dislocation can be the opportunity for self-evaluation which can bring (hopefully) an expansion of mental, moral, and spiritual horizons.

This is also true of present shame reaching back into the past. Aquinas is conscious of family shame for a criminal forbearer (*ST* I-II. 81.1 *ad* 6) as an analogy for original sin. On a broader scale, Probyn points to the socially transforming aspect of shame and regret together with their relations to collective responsibility and reconciliation concerning indigenous peoples, especially in relation to injustices from previous generations.[77] Through shame, the moral horizon expands beyond an individual focus on guilt to one that is more communal and shared. Shame can reach back into the past to those who have been victims of injustice and can carry moral weight into the present.[78]

Finally, we can we take a lead from René Girard and ask whether the form of shame associated with a concern for victims is "the secular face of Christian love"? (Girard 2001, 161). Is it possible to consider shame as a cultural response that suggests a movement in self-transcendence, a further differentiation of consciousness in which perception of truth is broadened and responsiveness to value is enhanced? Shame is now intertwined with what Anthony Kelly suggests is an unprecedented "stirring of conscience" that is the transforming effect of one particular 'risen' victim. It is a sign that the "paschal mystery of Christ's death and resurrection is, in fact, penetrating human history in a surprising way" (Kelly 2007, 5). We seem to have uncovered another 'face' to shame.

End Notes

1. Some of the themes of the following discussion are explored in Ryan (2008).
2. Calhoun reminds us that insofar as a virtue ethic invites us to identify our good with the common good, then how others perceive us (and our being sensitive to such perception) are important because we need to be available as predictable and reliable members of a shared moral community.
3. Thomas Aquinas treats of shame as a foundational moral response in *ST* I-II.24.4, as one of the six species of fear in *ST* I-II.41.4 and as an integral part of the virtue of temperance in *ST* II-II.144. He also has treatments in his *Commentary on the Nichomachean Ethics* Book 4, 17 a-m *et passim.* For translations of the *Summa*, the author has consulted the Latin/English (Blackfriars) version of the English Dominican Province (London: Eyre and Spottiswoode, 1963–1975), the *Summa Theologica of* St. Thomas Aquinas, 2nd rev. ed. 1920, trans. Fathers of the English Dominican Province, in the on-line version www.newadvent.org/summa/ and the new translation by Alfred J Freddoso, on-line version at http://www.nd.edu/~afreddos/summa-translation/TOC.htm, accessed 20/12/2008. Unless indicated, translations are from the Blackfriars' version. Summaries or paraphrases are the author's.

4. These contrasting forms of 'historical' and 'classical' consciousness are used by theologian Bernard Lonergan, SJ. See Lonergan (1967, 126–133).
5. See his discussion on the phenomenology of love in *ST* I-II.28.5. Elsewhere, concerning his discussion of friendship (*ST* II-II.23.1), J-P. Torrell notes Aquinas' "delicate sensibility" and that it is "difficult to think that the man who spoke in this way had nothing but a literary knowledge of affection." (Torrell 1996, 283).
6. This is specifically the case with the second part of the *Summa Theologiae.*
7. Probyn discusses the differing usages across contemporary disciplines. Silvan Tomkins, she notes, uses the language of affect which reflect his link with the scientific tradition and his understanding that 'affects' are innate to organisms. Alternatively, Probyn notes that 'emotion' tends to be used "by those who insist that emotion is social and cultural in genesis." She suggests, as an apt (and working) description that "emotion refers to the social expression of affect, and affect in turn is the biological and physiological experience of it." (Probyn 2005, 25).
8. For these dynamics see *ST* I-II.22.1-3.Naturally, a specific emotional response can be modified or may not occur if the apprehension or belief is mistaken or awareness of the object through memory fades. See Aquinas on fear *ST* I-II.42.2 and on anger *ST* I-II.48.2.
9. These same emotions, in Aquinas' view, can be found in the spiritual, non-bodily dimension of human existence (intellective appetite (*appetitus intellectivus*) and movements of the will called *affectus*) and may or may not have reverberations at the bodily level. Aquinas speaks of the *affectus simplex* where 'simple' contrasts with *affectus* properly speaking. *Affectus simplex* connotes the roused activity of the will (*appetitus intellectivus*) that occurs without any physical change or disturbance of the soul (*absque passione vel animi concitatione*), for instance, in desire (*concupiscentia*) for wisdom or hatred (*odium*) of vice. In this sense, it can also be applied to God or the angels (*ST* I.82.5 *ad* 1).
10. Probyn explores this further in relation to the feeling of being an outsider and 'out of place' in the experience of not 'getting a joke' or not recognising something as part of local lore and practice prompting the comment "You're not from here." (Probyn 2005, 38).
11. Shame is "a kind of primal reaction to the very possibility of love—either of oneself or of another" (Probyn 2005, 3).
12. "You have no shame on your skin, you are crazy" (Probyn 2005, 33, citing a comment of a local of Mt. Hagen, Papua New Guinea).
13. The 'shame-guilt' question reflects a contrast underlying Probyn's work which will be relevant to our later discussion, namely between the collectivist (shame-based) culture compared to the individualist (guilt- based) culture. A shame-culture is collectivist in that "persons understand themselves as parts of groups or collectives such as family, tribe or nation" (Triandis 1995, 2). They are defined by those groups and do not understand themselves as having a 'separate identity.' In contrast with an individualist or 'guilt-culture', members are motivated by "group norms rather than individual needs or aspirations." (Rohrbaugh

2002, 30) citing Triandis (1995). Hiebert gives a more specific picture when he notes that "in a shame-culture (sometimes referred to as "honour-shame culture"), what other people believe is much more powerful. Indeed, my principles may be derived from the desire to preserve my honour or avoid shame to the exclusion of all else." (Hiebert 1985, 212).

14. Interestingly, we can feel shame when others incorrectly perceive us as engaging in unworthy action. But even in those cases we have a deeper interest in being good as opposed to merely seeming to be good. Also, despite the social nature of shame, there is truth in saying that what shames you may not shame me.
15. We can also feel shame for others, particularly our children. Moreover, it must be remembered that, apart from passing moments of 'fragility', I can have an interest in wanting and thus willing myself to have a sense of shame, which may underpin the 'temporary' dimension noted here. Again, I might want to educate myself to have a greater sense of shame and thus shame can be cultivated by a voluntary examination of self.
16. By 'second nature' Bourdieu is suggesting two things: first, a comparison with humanity's 'first nature'—namely the basic makeup of human being understood universally; second, the practices and habits of being human that we use without thought or effort, namely, 'naturally,' e.g., walking. Bourdieu's use of *habitus* must be distinguished from *habitus* as found in Aristotle (*hexis*) and developed by Aquinas in his moral theory. This will be discussed later but a working definition suffices here. A habit (*habitus*) is a quality that is difficult to change, adding to nature (and its inclinations) through repetition thus giving it ease in performance. Virtues are good habits of living or conduct in that they nourish and enhance authentic humanity (according to 'right reason'). They are moral practices that are 'second nature' to a virtuous person. Perhaps by analogy with virtue theory, Bourdieu's use of 'second nature' refers to those aspects of history, culture and biography that shape us so profoundly they are 'forgotten', namely we do them with such ease and facility it is as if we were born that way. For a basic discussion of habit and virtue, see Fagothey (1986, 199–212).
17. Graham Ward citing George Steiner's discussion of the increasing banality [divorce of language and reality] with respect to the new pornographers "who parade the vital privacies of sexual experience, taking away the words that were spoken in the night to shout them from midmorning rooftops." (Ward 2002, 159). Ward cites Steiner (1967, 40).
18. This does not suggest that one is reflectively and consciously aware at the time that one is experiencing shame. Evaluation certainly involves an acceptance of the legitimacy of one's feeling shame (consciously or not) through acknowledging the temporary 'fragility' it brings. It may only mean, at a basic level, that one may feel shame (blush), draw back, and because that blush was unpleasant, avoid the situation in future.
19. We noted earlier some brief comments in *habitus* in Aquinas building on Aristotle (*hexis*). Aquinas has a compact discussion of *habitus* in *ST* I-II.49–54. While there are differing interpretations of the precise meaning of the term, Cessario notes that it is usually translated as a 'state of character' and gives reference to *Nichomachean Ethics* Bk.2, Ch. 6 (1106b36). See Cessario (1991, 34).

20. Aquinas has an extensive discussion of shame in relation to these aspects of human life in *ST* II-II.141–154. Further references will be given in the discussion later in this chapter.
21. The phrase he uses for 'of its very nature' is *ex sua specie* or *secundum speciem suam*. Earlier, in *ST* I-II.24.1, Aquinas argues that emotions, considered in themselves, i.e., intrinsically (*secundum se*), as natural phenomena or psychological facts, cannot be called morally good or evil.
22. Given the spontaneous nature of emotional responses, it is not uncommon to see them described as psychological facts that are 'morally neutral.' This is understandable particularly in relation to what are often referred to as the 'negative' emotions (those that make us feel uncomfortable, such as anger or fear). The danger is that because we feel 'bad' (equilibrium is disturbed) we conclude that we are 'bad' morally (we have done something wrong). A full discussion of this is beyond the scope of this chapter. Suffice it to say that, for Aquinas, we do have some level of responsibility for our emotions and our emotional life. Emotions can be morally significant in themselves and not just from our attitude to them. Without our emotions, we cannot be fully authentic human beings made in the image of God, in that we cannot *be affected by and respond in* the world of relationships. Hence, we need the affective virtues. See Harak (1993) and Murphy (1999).
23. One must keep in mind the distinction between acts such as sneezing or shuffling one's feet, referred to as 'acts of man' (*actus hominis*), and 'the human act' (*actus humanus*) which is morally significant since it is informed by human rationality—namely, it is deliberate and free. Aquinas' treatment of the moral act in terms of object, end, and circumstances is in *ST*.I-II.7.18.
24. *Ethics* II.7. 1108a32. In the same place Aristotle says that *verecundia* is not a virtue.
25. It is beyond our concern here but it is worth noting that, in this article, Aquinas proposes that an emotion that is evil by its very nature is envy (*invidia*) understood as "chagrin over another person's good fortune" since its object is "at odds" with "right reason."
26. This is rendition in the Blackfriars version (Vol 19, 1967, 43) compared with 'base fear' used in the earlier 1920 version on the New Advent website. See above n. 3.
27. This is clarified in the same article when Aquinas says that those emotions are good which 'create a favourable attitude towards something truly good or an unfavourable one towards something really evil; and those emotions are evil which create an unfavourable attitude towards something truly good, or a favourable one towards something really evil' (*ST* I-II.24.4 *ad* 2).
28. Green offers a helpful analytical tool for this discussion. Like any emotion, shame is an intentional mental state. It must be shame about something, hence have an "intentional object, or target." The person must believe that shame's object has a property that makes it inimical to his/her good or happiness. This belief is the 'motivating belief' and the property of the object is the 'motivating property.' Once this belief is formed, there occurs in the subject a "salient bodily change" that breaks into or suffuses consciousness "as *feeling* or *affect*."

The intentional object of the belief that occasions shame is base behaviour and its accompanying disgrace that moves a person to desire to "shrink or hide." This is what differentiates shame from hatred (desire to destroy its object) or anger (the desire to resist its object). See Green (2007, 115–116). For how one might understand 'as *feeling* or *affect*' see Probyn's discussion earlier n. 7.

29. Cited by Probyn (2005, 120) from Nussbaum (2001, 23–24).
30. Aquinas distinguishes the 'real' mean in 'operative' virtues (those which result primarily in action). For instance, with justice there is an objective benchmark that determines the measure of the just action. But with those moral virtues whose primary outcome is 'immanent' (affecting the subject and the capacity to respond emotionally as in temperance), the mean cannot be quantified in the same way. Hence, he uses the term 'rational' or 'reasoned' mean. Subjective factors such as stage of development, particular gifts, strengths, context, and so on, all contribute to the prudential judgment of the right balance of emotion, and, where appropriate, of action by *this* person, in *this* situation. See *ST* I-II.64.1 and 2.
31. This needs to be understood in the broader context of Aquinas' discussion. He sees the attractiveness of temperance in terms of the beauty (and moral goodness—*honestas*) inherent in a harmonious and balanced body/spirit existence. See *ST* II-II.141.2 *ad* 3 and 8 *ad* 1.
32. See *ST* II-II.144.
33. This is consistent with Aristotle's view that appropriate self-regard (*philautia*) is integral to human flourishing. Shame is entailed in self-care as moral sensitivity to actions that could reflect, or have reflected, badly on oneself (and a sense of remorse and even a desire to atone). See Oakley (1992, 74). Nussbaum notes that shame "requires self-regard as its essential backdrop. It is only because one expects oneself to have worth or even perfection that one will shrink from or cover the evidence of one's nonworth or imperfection" (Nussbaum 2001, 196).
34. Aquinas cites the carefully worded phrase from Aristotle. Shame is present in the virtuous person not absolutely but conditionally or hypothetically ('on this hypothesis'), namely, in the sense, that someone good would be ashamed *if* they were to do this or that…' *Ethics* IV, 9, 1128b21, Lect. 17.
35. See n. 17 above.
36. Clearly, there are deviations from the 'normal.' A person may find the feeling of righteous anger as quite comforting. Alternatively, shame, fear and anger can be free-floating and, at times, can seek out and 'hook onto' an object through the psychological mechanism of projection.
37. Probyn refers to legal initiatives (e.g., community or 'circle' sentencing) taken in New Zealand concerning Maori offenders and in Australia concerning indigenous peoples. In close communities, shaming the offender works better than formal sanctions (retributive justice) since individuals care deeply about what family and friends think about them. It also brings home to offenders the consequences of their actions when confronted by those they have harmed in some way (Probyn 2005, 90–98). Such processes are reminders how cultures can learn from one another (as here individualist/guilt from collectivist/shame cultures).

They alert us to the danger, perhaps especially for westerners, of seeing collectivist/shame-based cultures as being essentially conformist with a moral system centered on what other people expect a person to do. One wonders whether beneath this is an implication (maybe questionable?) that an individualist/guilt-based framework of socialization is better able to help its members cultivate values that are personally appropriated such that they act from deep conviction.

38. See n. 13. Shame and 'face' are central in Asian cultures. 'Face' is about how others perceive a person's value, status, credibility or social standing. The Chinese terms for face are '*lian*' and '*mianz*'. This has a significant bearing on business and organizational life, especially for westerners dealing with peers from such cultures. A key consideration is that "the individualism of the West involves a conception of the self as an autonomous, independent, and unique person" whereas, "in collectivist cultures of Asia, the core cultural norm and ideal is to achieve and foster harmonious interdependence among group members." (Kim and Nam 1998, 526). Again, it is interesting how even cross-culturally the bodily expression of shame (respect) is associated with the 'face'—'loss of face,' keeping face,' etc. *Blush* is the title of Probyn's book for the 'many faces of shame.' Citing Gregory of Nyssa, Aquinas uses *erubescentia* (red-faced, blushing) of someone who feels shame at the prospect of doing something morally disgraceful (against personally appropriated values). This contrasts with 'being ashamed' (*verecundia*) as when a person does something 'disgraceful' but does not want to be seen for fear of reproach (*ST* II-II.144.2).
39. Lester notes the many efforts to apply shame and guilt to Ruth Benedict's classification of societies (Benedict 1946). Using her classification, "many non-industrial societies and many modern Asian cultures [are] viewed as shame-cultures and Western cultures as guilt-cultures." In Lester's study of the cultural acceptability of suicide, after acknowledging that Benedict's classification is "an overly simple generalization", he says that "shame-motivated suicide is common in some Asian societies and is often found in their historical accounts and myths." He continues, "despite the fact that Confucianism and Buddhism both condemn suicide, the Chinese have typically viewed suicide committed out of loyalty to the family or to "save face" as acceptable." Finally, Lester points out that "suicide committed as a result of shame is most commonly associated with Japan where, as Davidson and Schaffner (1977) noted, child-rearing techniques utilize shame in order to socialize and enculturate the child." (Lester 1997, 358, citing Davidson and Schaffner 1977). Further comments in this classification *infra*.
40. The words 'damaged' and 'lethal' have contemporary relevance on a broader scale. For instance, in India and other parts of Asia, cultural factors mean that incurring financial debt can cause more than personal upset, loss, and bankruptcy. In 2010, there was a significant rise in people committing suicide over the shame of not being able to repay debts in micro-banking loans. See 'India's Microfinance Industry Fuels Mass Suicides', online at http://foreclosureblues.wordpress.com/2011/01/04/, accessed 26 March 2011. Also Buncombe (2010).

41. The strategies needed to dismantle and re-establish the deeply—embedded cultural attitudes and *habitus* underlying this 'attunement' involve resources and an investigation beyond the scope of this discussion.
42. Some comments on the various translations of *honestas* will be given later.
43. See Probyn (2005), in particular Chap. 3 'The Shamer and the Shamed' and Chapter 4 'Ancestral Shame.'
44. Gilby (1968), Notes, Trans. of *Summa Theologiae* (*ST* II-II.141–154), Vol. 43, 1968, 55.
45. As has been noted earlier, for Probyn, "shame goes further…to how we think about ourselves" and often demands a global "re-evaluation of the self" (Probyn 2005, 45). In that sense, she argues, shame "lingers deep within the self" (Probyn 2005, 2, 45–6). Stocker and Hegeman pose the question as "whether the shame is *contained* or *localized* or instead is *globalized and spread throughout the entire self*." (Stocker and Hegeman 1996, 222) (emphasis in original).
46. The actual text is "Nevertheless, as with other emotions, some feeling of shame may forestall the stand he makes on grounds of reason."
47. This phenomenon involving shame and loss of face has contemporary relevance, if even in a more extreme form. Lester notes that "when presented with various motives for suicide, 17 % of the Japanese people surveyed considered 'when a person loses face' as an acceptable motive." (Lester 1997, 359), citing P. Lewin, 'The Japanese life-plan and some of its discontents,' *Hiroshima Forum for Psychology* II (1986), 39–56. Again, Wei and Chua note the important role 'life stressors' played in precipitating suicide in studies done within China and India. From amongst these 'stressors', factors such as marital problems, other family difficulties, job loss, and financial setbacks (such as bankruptcy) emerged as "significant proximal causes" or "important precursors" to suicide (Wei and Chua 2008, 435).
48. This distinction of 'disgrace' and 'discretion' shame is suggested by Schneider. He relates them to two words in French: 'honte' which is Aquinas' 'shamefacedness' (*erubescentia*) = 'discretion-shame' and 'pudeur' which is equivalent to Aquinas' 'shame' (*verecundia*) = 'disgrace-shame.' See Schneider (1977, 18–19).
49. See n. 34.
50. The article builds on the *sed contra* concerning Aristotle 'holding that a man feels shame the more from those he is continually with,' and citing *Rhetoric* II, 6, 1384a36. Aristotle's reasons for thinking friendship is needed to cultivate each other's virtue and to help, through example and human interchange, in leading a flourishing life are outlined in *Nichomachean Ethics* 1169b8-1170b25. See also Aquinas' commentary on Books 8 and 9 of the same work dealing with friendship in his *Sententia libri Ethicorum*, available online at http://www.thomasinstituut.org.
51. The body of the article ends with 'The witness may speak from closer knowledge of the facts of our case, and here our associates know us best, whereas we set less store on the good opinion of passers-by and strangers.' A similar comment is made in the reply to the first objection.

52. Aquinas' treatment in this article is a more developed form of that found in Aristotle's *Rhetoric* II. 6. Cua notes Aristotle's more extended treatment of shame in his *Rhetoric* compared to the *Nichomachean Ethics*. Aristotle has a greater appreciation that the experience of shame is 'ethically significant because it is an expression of an aspect of moral consciousness' and 'implicitly entails the acceptance of certain standards for interpersonal relationships.' See Cua (2007, 152–53).
53. For instance, Tomkins (1995, 1963).
54. She tentatively notes that her position suggests that "*Essentialist* or *ethnocentric* epithets hover in the air" (Probyn 2005, 28, italics in original).
55. Leget notes this in comparing Aquinas' account of the emotions with Martha Nussbaum's *Upheavals of Thought* where she engages with contemporary psychological theories (e.g., Martin Seligman, Richard Lazarus, Antonio Damasio), especially concerning a greater appreciation of emotions as intentional and embodied "cognitive value-laden appraisals" (as they are seen by Aquinas too). Again, Leget notes that there are "affinities between Aquinas and contemporary research as concerns overcoming mind/body dualism and the emotion/reason distinction." (Leget 2004, 576). See above n. 29 for Nussbaum reference.
56. For instance, in *ST* I-II.38.1-5, one finds a five-step strategy to deal with sadness, depression, loss, and grief. This parallels the steps suggested by James and Evelyn Whitehead in dealing with 'negative' emotions, namely, to name, claim, tame, and aim one's emotions. See Whitehead and Whitehead (1994, 175ff). Victor White says of Q. 38 on the remedies of depression, "we find a surprisingly up-to-date application of the principle of functional opposition and compensation, recognition of both the organic and the psychological function of weeping, an exact description and explanation of the releasing effect of transference through 'a certain imagination that others bear the sufferer's burden' ('*quaedam imaginatio quod onus alii cum ipso ferant*'), and more than a hint of such 'modern' methods as hydrotherapy and prolonged narcosis." (White 1964, 103).
57. See nn. 13, 38, 39, 40.
58. *The Australian Concise Oxford Dictionary* (1987), 511.
59. Honour as 'recognition' and appreciation is due to moral excellence. He says that 'the honourable amounts to the same as being virtuous.' (*ST* II-II.145.1).
60. Stocker and Hegeman (1996, 291). The desire for recognition of virtue entails the desire to live with a community of those who recognize and appreciate 'the virtues of common pursuit' (in Nancy Sherman's phrase) which is another expression of Calhoun's participation in 'shared practices.' These are people who, for Stocker and Hegeman, "…are alive to and appreciate each other, including what each contributes; and who, as part of living within such a community, give thanks and recognition, assurance and mutual support." *Ibid*. 291–92. Also see Sherman (1993, 277–299).
61. Naturally, 'change and development' is evident in the contemporary phenomenon of globalization and an increasing cross-cultural interaction. This has prompted a certain social tension between individualist/western cultural

attitudes when they are imbibed and at work within collectivist/eastern cultures. The reverse is true when there is an increasing proportion of immigrants from community/shame-based cultures (e.g., from Asia, Africa, Middle East) living in countries such as the United States and Australia. The repercussions of this issue are evident on a global scale in the relation of Islam to Western democracies.

62. Morris (1987) and Walker Bynum (1984).
63. Walker Bynum qualifies this in her reply to Morris. She prefers to speak of a 'discovery' of the 'self' (rather than the individual) first, as a subject of inner mystery, second, concerning a wider range of choices about 'callings' and social roles and finally, how this 'self' related to groups, society and the Church.
64. Many Eastern, African, Polynesian, and Asian societies are still kin and clan based (collective cultures). For instance, in some of these cultures (e.g., Chinese), when politely addressing strangers, it is customary to refer to older individuals as 'auntie' or 'uncle', thus subsuming social encounters under the rubric of the family.
65. A contemporary parallel is developed in Gaita (1998).
66. See Hall (1995, 76–77).
67. In the style of Aquinas, as a *sed contra*, perhaps the myth of Eros and Psyche tries to capture human yearnings that philosophy did not have the tools to articulate.
68. See *ST* II-II.23.1 *ad* 2. For a fuller discussion of this see Ryan (2010).
69. See Galatians 3: 28 "There is neither Jew nor Greek, there is neither slave nor free, there is no male and female, for you are all one in Christ Jesus."
70. See his comment "Truth, whatever its source, is of the Holy Spirit" (*ST* I-II.109.1 *ad* 1, citing St. Ambrose).
71. See n. 38.
72. See text and n. 2.
73. See Luke 6: 43–45 on the tree bearing good and bad fruits and image of good person drawing what is good 'from the store of goodness in his heart.'
74. See also Williams (1993).
75. Also Williams (1993, 93).
76. See n. 38.
77. This is further elaborated in Gaita (1999).
78. See Gaita (1999, 87–106).

References

Arbuckle, Gerald A. 2010. *Culture, inculturation and theologians: A postmodern critique*. Collegeville: Liturgical Press.

Aristotle. 1976. *The ethics of Aristotle: The Nichomachean ethics*. Trans. J.A.K. Thomson and Hugh Tredennick, Intro. J. Barnes. Harmondsworth: Penguin.

Auerbach, Erich. 1974. *Mimesis: The representation of reality in Western literature*. Princeton: Princeton University Press.
Aquinas, 1963–1975, Summa Theologica, Trans. Blackfriars version of the English Dominican Province. London: Eyre and Spottiswoode, 1963–1975.
Benedict, Ruth. 1946. *The chrysanthemum and the sword*. London: Routledge & Kegan Paul.
Buncombe, Andrew. 2010. Suicides and pressure tactics tarnish image of micro-finance lending. *The Independent*, 22 November. Online at: http://www.independent.co.uk/news. Accessed 26 Mar 2011.
Calhoun, Cheshire. 2004. An apology for moral shame. *The Journal of Political Philosophy* 12(2): 127–146.
Cessario, Romanus, O.P. 1991. *The moral virtues and theological ethics*. Notre Dame: University of Notre Dame Press.
Coleman, Gerald D., SS. 1992. *Human sexuality: An all-embracing gift*. New York: Alba House.
Cua, Anthony S. 2007. The ethical significance of shame: Insights of Aristotle and Xunzi. *Philosophy East and West* 52(3): 147–202.
Davidson, L., and B. Schaffner. 1977. Contrasts and similarities in the handling of emotions in two different cultures. *Hiroshima Forum for Psychology* 4: 59–67.
Fagothey, Austin. 1986. *Right and reason: Ethics in theory and practice*, 8th ed, Revised by Milton A. Gonsalves, 199–212. Columbus: Charles E Merrill.
Gaita, Raimond. 1998. *Romulus, my father*. Melbourne: Text Publishing.
Gaita, Raimond. 1999. *A common humanity: Thinking about love & truth & justice*. Melbourne: Text Publishing.
Gilby, Thomas, OP (trans.). 1968. Intro. & notes, Thomas Aquinas, *Summa Theologiae* (2a.2ae. qq.141–154), vol. 43. London: Eyre and Spottiswoode.
Girard, René. 2001. *I see Satan fall like lightening*. Maryknoll: Orbis.
Green, Keith. 2007. Aquinas's argument against self-hatred. *Journal of Religious Ethics* 35(1): 113–139.
Hall, Pamela M. 1995. *Narrative and the natural law: Interpreting Thomistic ethics*. Notre Dame: University of Notre Dame Press.
Harak, G.Simon. 1993. *Virtuous passions: The formation of Christian character*. Mahwah: Paulist.
Hiebert, Paul G. 1985. *Anthropological insights for missionaries*. Grand Rapids: Baker Book House.
Kelly, Anthony J., C.Ss.R. 2007. The resurrection and moral theology: Does it make any difference? *The Australian Ejournal of Theology* 10: 1–8.
Kerr, Fergus. 2002. *After Aquinas: Versions of Thomism*. Oxford: Blackwells.
Kim, Joo Yup, and Sang Hoon Nam. 1998. The concept and dynamics of face: Implications for organizational behavior in Asia. *Organization Science* 9(4): 522–534.
Leget, Carlo. 2004. Martha Nussbaum and Thomas Aquinas on the emotions. *Theologica Studies* 64(3): 558–580.
Lester, David. 1997. Shame and suicide in Asian countries. *Suicide and Life-Threatening Behaviour* 27(4): 352–361.
Lonergan, Bernard S.J. 1967. The transition from a classicist world-view to historical mindedness. In *Law for liberty*, ed. James E. Biechler, 126–133. Baltimore: Helicon Press.
Morris, Colin. 1987. *The discovery of the individual*. London: SPCK.
Murphy, Claudia Eisen. 1999. Aquinas on our responsibility for our emotions. *Medieval Philosophy and Theology* 8(2): 163–205.
Nussbaum, Martha C. 2001. *Upheavals in thought: The intelligence of the emotions*. Cambridge: Cambridge University Press.
Oakley, Justin. 1992. *Morality and the emotions*. London/New York: Routledge.
Ormerod, Neil. 2007. In defence of natural theology: Bringing god into the public realm. *Irish Theological Quarterly* 72: 227–242.
Probyn, Elspeth. 2005. *Blush: Faces of shame*. Sydney: University of New South Wales Press.
Putnam, Daniel. 1995. Defence of Aristotelian honour. *Philosophy* 70(272): 286–288.

Rohrbaugh, Richard L. 2002. Ethnocentrism and historical questions about Jesus. In *The social setting of Jesus and the Gospels*, ed. Wolfgang Stegemann et al., 27–43. Minneapolis: Fortress Press.
Ryan, Tom. 2008. Healthy shame? An interchange between Elspeth Probyn and Thomas Aquinas. *Australian eJournal of Theology* 12: 1–18.
Ryan, Tom. 2010. Aristotle on compassion: Has he anything to offer today? *Irish Theological Quarterly* 75(2): 157–174.
Schneider, Carl D. 1977. *Shame, exposure and privacy*. Boston: Beacon.
Sherman, Nancy. 1993. The virtues of common pursuit. *Philosophy and Phenomenological Research* 53: 277–299.
Steiner, George. 1967. *Language and silence*. London: Faber.
Stocker, Michael, and Elizabeth Hegeman. 1996. *Valuing emotions*. Cambridge: Cambridge University Press.
Stump, Eleonore. 2005. *Aquinas*. New York: Routledge.
The Australian Concise Oxford Dictionary. 1987. Melbourne: Oxford University Press.
Tombs, David. 1995. "Shame" as a neglected value in schooling. *Journal of Philosophy of Education* 29(1): 23–32.
Tomkins, Silvan S. 1963. *Affect, imagery, consciousness, vol. 2, The negative affects*. New York: Springer.
Tomkins, Silvan S. 1995. Shame-humiliation and contempt-disgust. In *Shame and its sisters: A Silvan Tomkins reader*, ed. Eve Kosofsky Sedgwick and Adam Frank. Durham: Duke University Press.
Torrell, J-P., O.P. 1996. *Saint Thomas Aquinas: Vol. 1: The person and his work*. Washington, DC: Catholic University of America Press.
Triandis, Harry C. 1995. *Individualism and collectivism. New directions in social psychology*. San Francisco: Westview.
Walker Bynum, Caroline. 1984. Did the twelfth century discover the individual. In her *Jesus as mother: Studies in spirituality of the high middle ages*, 82–109. Berkeley/Los Angeles: University of California Press.
Ward, Graham. 2002. In the daylight forever?: Language and silence. In *Silence and the word: Negative theology and incarnation*, ed. Oliver Davies and Denys Turner. Cambridge: Cambridge University Press.
Wei, K.C., and H.C. Chua. 2008. Suicide in Asia. *International Review of Psychiatry* 20(5): 434–440.
White, Victor, O.P. 1964. *God and the unconscious*. Glasgow: Fontana Books.
Whitehead, James D., and Evelyn E. Whitehead. 1994. *Shadows of the heart: A spirituality of the negative emotions*. New York: Crossroad.
Williams, Bernard. 1993. *Shame and necessity*. Berkeley: University of California Press.

Part II
Aquinas and the East: Comparative Approaches

Can Morality Be Taught? Aquinas and Mencius on Moral Education

Anh Tuan Nuyen

Question 11 of Aquinas' *De Veritatae* ('Disputed Questions on Truth') 'On the Teacher' (*De Magistro*) begins with the question: 'Can a Man or Only God Teach and Be Called a Teacher?'[1] On the surface, this question is rather puzzling. It seems rather odd to ask if any one person can teach and be called a teacher of others given the fact that the teaching of men and women by some other men and women has been going on for thousands of years and has reached the point where it is now a major enterprise in a modern society (and a booming industry in many economies!). How can Aquinas' question be understood given this history? While Aquinas discusses education in general, I will argue in this paper that we can make better sense of Question 11 if we understand it as a question about moral education. Understood as such, this question is not at all puzzling—it is indeed pertinent—given the fact that moral skeptics certainly do not believe that morality can be taught and many believers in morality nevertheless think that one cannot be taught to be moral.[2] Understood as a question about the possibility of moral education, Aquinas' answer invites some interesting comparisons with the Confucian account of moral education, particularly that of Mencius. Drawing out these comparisons will be the main task in this paper.

1 Aquinas on Education

According to Aquinas, "certain seeds of knowledge pre-exist in us" in the form of complex and simple concepts of the understanding (Aquinas 1953, 82). A person "is said to acquire knowledge" when such person's mind applies these concepts to "particular things" (Aquinas 1953, 82). Knowledge, then, may be said to pre-exist in the learner but Aquinas stresses that it does so "not ... in the purely passive sense, but

A.T. Nuyen (✉)
Department of Philosophy, National University of Singapore, Singapore, Singapore
e-mail: phianhtn@nus.edu.sg

T.B. Mooney and M. Nowacki (eds.), *Aquinas, Education and the East*, Sophia Studies in Cross-cultural Philosophy of Traditions and Cultures 4,
DOI 10.1007/978-94-007-5261-0_6,

in the active sense" (Aquinas 1953, 83). The mind possesses an active power, which Aquinas calls 'natural reason,' that acts on the 'seeds' to turn them into knowledge. The pre-existing 'seeds' of knowledge can 'sprout' into knowledge in two ways: (1) when "natural reason by itself reaches knowledge of unknown things" and (2) "when someone else aids the learner's natural reason" (Aquinas 1953, 83). Aquinas calls the first way of learning *discovery* and the second way *learning by instruction* (Aquinas 1953, 83). The person who "aids the learner's natural reason" and helps the latter "learn by instruction" is the instructor, or the teacher. The teacher teaches by "manifest[ing] to [the learner] the reasoning process which he himself goes through by his own natural reason" (Aquinas 1953, 83).

To someone who asserts that only God can teach, Aquinas responds by making a distinction between 'interior' teaching and 'exterior' teaching and uses it to argue that when "Augustine proves that only God teaches, he does not intend to exclude man from teaching exteriorly, but intends to say that God alone teaches interiorly" (Aquinas 1953, 85). God is said to teach interiorly when He makes the "light of reason," which "is implanted in us by God," work in such a way that we can learn unaided, that is, learn by discovery (Aquinas 1953, 83–84). However, when a learner does not apply natural reason properly, or when natural reason somehow fails to work, a teacher can assist by "supplying external help to it to reach the perfection of knowledge" (Aquinas 1953, 85). Since the teacher does not 'endow' the learner's mind with the light of reason but merely 'co-operates' with it, the teacher is said to teach only 'exteriorly' (Aquinas 1953, 85). The distinction is bolstered by being compared with the distinction between natural healing and doctor-assisted healing. The natural light of reason is compared with the natural healing power of the body, which heals the body 'interiorly.' 'Exterior' healing occurs when a doctor "assists nature, which is the principal agent, by strengthening nature and prescribing medicines, which nature uses as instruments for healing" (Aquinas 1953, 82).

Aquinas' position is further elaborated in Articles 2–4 of the *De Magistro*. In Article 2, St. Thomas argues that since it is God who endows us with the light of reason, without which there can be no knowledge, no person can "be called his own teacher or be said to teach himself" even when he acquires knowledge without the aid of a teacher (Aquinas 1953, 89). When a person discovers things, by himself or herself as it were, he or she is in fact taught by God, interiorly. In Article 3, it is pointed out that while the teacher teaches exteriorly by 'manifesting' to the learner 'the reasoning process' and 'co-operating' with the learner's own natural reason, he or she does not alter the process in any way. By contrast, it is possible for an angel to do so even though it is God, not any angel, who endows us with reason. An angel can "strengthen the infused light [of reason] to make man see more perfectly" (Aquinas 1953, 95). When our "intellectual light … is brought in contact" with an angel's intellectual light, it will get "intensified" by it because the latter is "more perfect" (Aquinas 1953, 95). In this way, an angel can teach us in a way that is in-between God's interior teaching and a teacher's exterior teaching. In Article 4, Aquinas makes a distinction between the 'active' life' and the 'contemplative' life (Aquinas 1953, 99). The active life is concerned with "temporal affairs" and is "directed to the help of our neighbor" while the contemplative life is concerned with "the intelligible natures of things" and is aimed at "the consideration of truth" (Aquinas 1953, 99–100). Aquinas then claims that teaching "is more properly the work of the active than of the contemplative life" although it is "in some sense a function" of the latter (Aquinas 1953, 100). Thus teaching, while guided by the quest for truth and for understanding the intelligible natures of things, is more about helping our neighbors, or in the words of St. Gregory cited by Aquinas, more about doing the "spiritual works of mercy" (Aquinas 1953, 99).

It is clear from Question 11 of *De Veritatae* that Aquinas favors the natural way of learning through discovery. The teacher can help the learner 'learn by instruction', but he or she should only assist nature in the process. The teacher's art is to "imitate nature," operating "in the same way and through the same means as nature" (Aquinas 1953, 83). The use of agricultural metaphors, such as 'seeds' of knowledge, and medical examples, such as doctor-assisted healing, amply demonstrate Aquinas' naturalistic position. Whether by discovery or by instruction, learning is the natural process of nurturing by which the seeds of knowledge grow into the plants of knowledge. The teacher should merely assist by preparing the conditions for the learning process to develop naturally. Since the most natural way is learning through discovery, a good teacher is one who assists the learner in such a way that the learner is well set on the way of discovery.

It is clear also that in Question 11, Aquinas, in speaking of teaching and learning, is also speaking about education in general. However, a question arises as to whether his remarks are still applicable in the case of moral education, of learning to be moral. Textual evidence suggests that they do apply to moral education. As pointed out above, Aquinas regards teaching to be "more properly a work of the active than of the contemplative life" (Aquinas 1953, 100). Also, as we have seen, the active life is concerned with temporal affairs rather than with the intelligible natures of things. Thus, teaching has to do more with practical matters, with actions and behaviors, than with theoretical truths, even though knowledge of truths guides practical judgments. Practical matters include skills and moral behavior. For Aquinas, the important temporal affairs are the moral ones. Indeed, the end of the active life is to "help our neighbor" (Aquinas 1953, 100). Aquinas makes clear that he endorses St. Gregory's view, quoting him as saying that "the active life consists in giving bread to the hungry and in teaching the ignorant the word of wisdom" and adding that the "works of mercy are part of the active life …[and] teaching is counted among the spiritual work of mercy" (Aquinas 1953, 99). Thus, the business of the teacher is, in part, the moral education of the learner.

Reading Aquinas in this way, the question asked in Article 1, namely, "Can a Man or only God Teach and Be Called a Teacher?" becomes pressing. It seems straightforward enough, even though it may not be easy, to teach someone a skill, such as how to bake a cake or how to operate a machine, or to teach someone how to apply a concept to a thing so as to know that thing, or a rule to a problem so as to solve it. But how can I teach someone, or help my neighbor, to become moral? The moral skeptics would certainly deny that a person can teach another to be moral. Many who believe in morality believe it to be a personal, or subjective, matter, which cannot be taught. Indeed, some think that it is presumptuous to suppose that someone can be taught to be moral insofar as the teacher presumes to be morally superior to the learner (indeed, Aquinas himself states that "teaching implies the perfect activity of knowledge in the teacher or master" [Aquinas 1953, 89]). Even if morality is objective, it may still be argued that it cannot be taught. For instance, Kant says that while testimony is a legitimate source of knowledge, one should not take another person's moral judgments to be true without verifying them for oneself.[3] One plausible reading of Kant's position is that moral knowledge is a matter of discovery, not of learning by instruction, hence is not something that can be taught. Morality, many people would say, is a matter of encouragement or persuasion, either by empathy or by arguments; it is not something that can be taught.

Believing in morality, Aquinas does not have to respond to the moral skeptic's argument. Concerning the suggestion that it is presumptuous to teach others morality, it is open to Aquinas to argue that it is no more presumptuous than to teach, or try to teach, others a practical or theoretical skill. Teaching does imply "the perfect activity of knowledge in the teacher" but only concerning a specific subject matter, not perfection itself. To teach someone how to drive a car, the teacher must be a competent driver, but it does not follow that he or she is more competent than the learner in any other areas. Likewise, "teaching the ignorant the word of wisdom" (Aquinas 1953, 99), implies only that the teacher has greater wisdom on the specific matter at hand but not generally. Aquinas would probably agree that even the most simple-minded person has something to teach us.

As for Kant, it is true that Kant counsels us not to take another person's moral judgment on his or her authority alone but to examine it ourselves. However, this does not mean that another person's moral judgment is worthless. If the hearer does examine such judgment and verifies it with his or her own internal resources, it will still have performed the function of encouraging the hearer to arrive at a judgment, which, if true, will be part of the hearer's knowledge. Depending on how the hearer receives another person's moral judgment, he or she may still be said to have been instructed by it if knowledge results in the end.

In any case, Aquinas does not envisage education to be purely a process of transmitting information from the teacher to the learner. The latter's natural reason will have to be engaged and any knowledge will have to be what grows from a pre-planted seed. It is for these reasons, as textual evidence shows, that Aquinas holds that a person can teach another moral knowledge, to know how to act morally and how to be moral, which is the same thing as teaching the learner to act morally, or to be moral, given the fact that Aquinas subscribes to a version of internalism about moral motivation, that is, the view that knowledge of the good motivates good actions. On the learners' part, the need for learning to be moral arises from the fact that we are capable of determining the moral character of some actions but not all actions that contribute to our flourishing, and when we are unable to do so, we need to learn from someone who is endowed with wisdom on such matters.[4]

2 Aquinas and Mencius on Education: A Comparative Reading

The brief reading of Question 11 of Aquinas' *De Veritatae* above will serve as a basis for my comparison of Aquinas and Mencius in this section. Comparisons between Aquinas and Mencius are not new. In a fairly extensive comparative work, Lee H. Yearly compares and contrasts Mencius' and Aquinas' theories of virtue and conceptions of courage (Yearly 1990). Comparative efforts like this are useful even though, as Yearly admits, there are vast differences between Mencius and Aquinas. For instance, they differ on methodology ("Analytic procedures and tools are central to Aquinas but they usually remain peripheral to Mencius" [Yearly 1990, 4]); on cosmology (Aquinas believes in a God-Creator whose *aseity* (reality) "is not even a conceptual possibility" [Yearly 1990, 4] in Mencius); on the list of virtues (for Aquinas, they include the natural virtues of wisdom, justice, courage, and moderation, and the 'infused virtues' of faith, hope, and charity, and for Mencius, they include humanity (*ren*), propriety (*li*), righteousness (*yi*), wisdom (*zhi*) and others such as filial piety (*xiao*)); and most importantly on human nature (Aquinas believes that there is a mixture of good and bad elements in human nature while Mencius believes that human nature is wholly good).

Despite these differences we can find similarities in general as well as similarities in the differences themselves. Comparative efforts are useful in that they can lead to a deeper understanding of the compared thinkers and perhaps to a re-assessment of the common readings of their thoughts. Unfortunately, Yearly focuses on Mencius' and Aquinas' theories of virtue generally and on their conceptions of courage in particular, and has nothing to say on their views on education and moral education. Indeed, this latter type of comparative study has been largely neglected. In what follows, I will make some comparative remarks on, rather than a thorough study of, Mencius' and Aquinas' views on education generally and on moral education in particular.

Yearly correctly points out that a good comparative effort should identify similarities in differences as well as differences in similarities. We can begin then by noting some of the differences from which similarities can be drawn. Like Confucius and Aquinas, Mencius spent most of his life teaching others. For Mencius, education is the most important activity, not just for any person but also for the nation as a whole. To rule effectively, the ruler must provide the people with a good education (after ensuring that they are adequately fed and clothed). A good education, in turn, is seen by Mencius as aiming at promoting moral conduct. While Mencius thought that his pupils ought to be educated in many different areas (including music), the ultimate aim of education was to become a 'gentleman' (*junzi*) or a person who has cultivated virtues, such as humanity (*ren*), propriety (*li*), righteousness (*yi*), filial piety (*xiao*) and fidelity (*xin*) to a superior degree. An educated person will display the appropriate virtue in the various relationships: "between father and son, there should be affection; between sovereign and minister, righteousness; between husband and wife, attention to their separate functions; between old and young, a proper order; and between friends, fidelity." (*Mencius* 3A4)[5]

The principal aim of education then is to become a moral person. In terms of Aquinas' division into the contemplative and the active life, it would appear that Mencius privileges the active over the contemplative. Furthermore, for Mencius, the ultimate aim of education is social harmony, a notion that hardly plays any role in Aquinas' discussion of education in Question 11 of *De Veritatae*.[6] Indeed, in the interest of social harmony, the rulers themselves should be morally educated in order to govern effectively. Thus, Mencius would have agreed with the following remark by Confucius: "Lead the people with governmental measures and regulate them by law and punishments and they will avoid wrongdoing but have no sense of honour or shame. Lead them with virtue and regulate them by the rules of propriety (*li*) and they will have a sense of shame and moreover set themselves right." (*Analects* 2.3) For Aquinas, by contrast, since the ultimate aim of education is perfection, the first important step is to overcome original sin, to triumph over evil.

Moving away from the differences above and focusing on education itself, the first similarity between Mencius and Aquinas is the key epistemological claim that combines elements of innatism (the view that knowledge is imprinted on the mind, or innate in us) and empiricism (the view that the mind is a blank slate and knowledge has to be learnt through experience). Thus, just as Aquinas speaks of the 'seeds of knowledge' implanted in us by God, which presumably include the seeds of moral knowledge, Mencius speaks of the 'four sprouts' that are naturally embedded in a person's heart-mind (*xin*). However, while Aquinas takes these seeds to be general concepts and principles, Mencius takes them to be the psychological beginnings of the virtues. He identifies these sprouts as sympathy, shame, deference, and a sense of right and wrong. Mencius writes:

> A heart-mind that sympathizes is the sprout of humanity [*ren*]; a heart-mind that is aware of shame is the sprout of rightness [*yi*]; a heart-mind that defers to others is the sprout of ritual propriety [*li*]; a heart-mind that approves and condemns is the sprout of wisdom [*zhi*]. (*Mencius* 2A6)

With the four 'sprouts' embedded in oneself, a person can discover moral knowledge unaided, as if being taught 'interiorly' by God: "If anyone having the four sprouts within himself knows how to develop them to the full, it is like fire catching alight, or a spring as it first bursts through." (*Mencius* 2A6) However, Mencius immediately qualifies this by saying: "If able to develop [the sprouts], he is able to protect the entire world; if unable, he is unable to serve even his parents," thus acknowledging that a person may fail to learn 'interiorly,' in which case he or she needs to be taught 'exteriorly' by a teacher. (*Mencius* 2A6)

As we have seen, Aquinas privileges learning by discovery over learning by instruction. In the same way, Mencius stresses the importance of 'interior' learning, of seeking within oneself, over being taught by others. In speaking of 'self-cultivation,' Mencius means not just the cultivation *of* the self but also the cultivation *by* the self. The cultivation of the self by the self is the more natural process. For Aquinas, it is like the body healing itself by its own internal healing power, which is better than being healed by a doctor. Making the same point, Mencius nevertheless prefers to extend the agricultural metaphor: "Take the barley for example. Sow the seeds and cover them with soil.... The plants shoot up and by the summer solstice they all ripen." (*Mencius* 6A7) All that is needed is the effort in sowing the seeds and watering the plants, and perhaps pulling out the weeds. The plants will grow on their own without any intervention. Indeed, Mencius tells the story of a farmer who pulls on the sprouts to make them grow but only succeeds in killing them: "Be not like the man of Sung ... who pulled on his grain sprouts because he was worried they would not grow." (*Mencius* 2A2) Unfortunately, the agricultural metaphor only goes so far. When the sprouts fail to grow, it could well be that the farmer is to blame, but it could also be the case that there is something wrong with the sprouts themselves. The analogy breaks down in the latter case insofar as Mencius does not accept that there could be anything wrong with the sprouts in the human heart-mind (*xin*). If these sprouts fail to grow, it is the agent himself or herself that is to blame, like the farmer of Sung, in which case exterior teaching and learning become necessary. The farmer of Sung needs to be taught how to cultivate rice.

How great is the need for exterior teaching? It would appear that Mencius sees much less a need for it than Aquinas. This difference follows from the fact that, as noted above, while Mencius takes human nature to be wholly good, Aquinas takes it to contain a mixture of good and bad elements, although given his subscription to the doctrine of original sin, it is probably fair to say that, for Aquinas, human nature is bad on balance. Despite his belief in the goodness of human nature, Mencius does acknowledge that some people often do bad things or fail to do good things. Farmers typically do not pull on their grain sprouts but some do, like the farmer of Sung. Indeed, at 1A7, Mencius speaks of King Xuan who knows that his people are suffering, that he ought to relieve their suffering, and yet fails to do so. External teaching is required when a person fails to cultivate the sprouts within and at 1A7 Mencius teaches King Xuan how to motivate himself to act benevolently towards his people. Still, Mencius believes that the sprouts that nature has placed in the human heart-mind (*xin*) are all good and strong and will naturally grow properly, given the right cultivation. There is no need to pull on the sprouts to help them grow: "Few are those in the world who do not help their sprouts grow [by pulling on them]." (*Mencius* 2A2) There is no great need for exterior teaching. By contrast, believing in the doctrine of original sin, Aquinas sees a greater tendency in humans to be bad, hence a greater need for moral education. As Yearly points out, "Aquinas believes inclinations to virtuous states remain but they also co-exist with, and even connect to, powerful inclinations to bad states" (Yearly 1990, 89). The latter 'powerful inclinations' manifest themselves in the seven deadly sins, although "Aquinas focuses his attention not on crude sins, like lust and gluttony, but on more complex and subtle sins, like envy, vanity, and spiritual apathy" (Yearly 1990, 89). Mencius, on the other hand, "in his more theoretical accounts of people's failures, tends to concentrate on inclinations that arise largely from people's bodily states" (Yearly 1990, 91). For Aquinas, human failures are due to "deformations that engage higher human faculties" while for Mencius, they are due to "deformations that arise from lower capacities" (Yearly 1990, 95). These differences point to differences not just in the urgency of exterior teaching as we have seen but also in the kind of teaching required, as we will see.

Before examining these differences further, it is worth re-emphasizing the striking similarity in Mencius' and Aquinas' naturalism. As we have seen, Aquinas and Mencius put considerable trust in nature and the natural way. Teaching itself must be natural and must follow the natural way; the teacher's art has to 'imitate nature.' Aquinas uses the example of the doctor healing a sick person to illustrate his point: in healing, the doctor can and should do no more than assisting the patient's own internal healing process. Mencius, on the other hand, relies heavily on an agricultural metaphor. However, commenting on Aquinas' philosophy of education, John Donohue employs precisely the agricultural metaphor, arguing that for Aquinas, the teacher is like the farmer who simply assists a plant to grow by watering, pruning and weeding (Donohue 1968). Since learning by discovery, which is wholly natural, is for Aquinas preferable to learning by instruction, a teacher should teach in such a way as to make the learner's experience as close as possible to that of learning by discovery. This is precisely the effect that Mencius achieves in his teaching of King Xuan how to discover within himself the natural compassion to care for his suffering subjects. As is evident from 1A7 of the *Mencius,* King Xuan feels that he has discovered the true nature of his feelings and dispositions, after his conversation with Mencius.

Returning to the difference between Aquinas and Mencius on human nature and its implications for teaching, it is useful to examine the difference between Mencius and Xunzi. Unlike Mencius, Xunzi believes that human nature is bad. Goodness is attained only through the deliberate exertion of artificial effort; it is not a matter of natural growth.[7] According to Xunzi, the 'sprouts' in a person's heart-mind are the fondness for profit and the feelings of hate and dislike. The natural desires are the base bodily desires. If these are allowed to grow without checks, they will grow into cruelty, villainy and lasciviousness. The people must be taught 'exteriorly' the moral rules, some invented and some discovered by the sage kings. Indeed, education alone will not be sufficient: artificial constraints must be imposed in order to ensure order and harmony. In a language that reminds us of Kant, Xunzi compares human nature with crooked timber and blunt metal, which will require artificial efforts to straighten and sharpen. Thus, there is no prospect of learning by discovery or 'interior' learning and the role of the teacher is absolutely crucial.

It is not suggested here that Aquinas' view on human nature is anywhere near Xunzi's. Aquinas does not believe that human nature is simply bad. Indeed, Xunzi's position is vulnerable to the objection that if human nature is bad, and there is no natural goodness in us, then how it is possible for some humans to be good enough to teach others, let alone be sage kings. There are no resources in Xunzi's account to answer satisfactorily Kant's famous question, namely "Can anything straight be constructed from the crooked timber of humanity?" (Kant 1963, prop. 6) Nevertheless, Aquinas' subscription to original sin and his stress on the 'inclinations to bad states' put him a little closer to Xunzi than to Mencius. It follows that Aquinas must take a harder line on moral education than Mencius. This is not evident in

Question 11 of *De Veritatae.* However, his medical metaphor is revealing enough. Thus, by comparing the teacher with the doctor, or teaching with healing, Aquinas leaves himself open to the suggestion that teaching can be firm at times, just as healing at times requires strong and bitter medicine. Following nature does not rule out firm teaching any more than following nature in healing rules out strong medicine. "Deformations that engage higher human faculties," to use Yearly's words, are a much more serious disease than "deformations that arise from lower capacities" as in Mencius' diagnosis (Yearly 1990, 95).

Given his diagnosis, Mencius is entitled to and clearly does take a soft line on moral education. For him, teaching is more like guiding, encouraging, and setting examples, than standing over and instructing. The agricultural metaphor employed by Mencius is perfectly appropriate as a developmental model for the cultivation of the self. In the best farming practice, plants are allowed to grow naturally. Weeding and killing pests may sound harsh, even violent, but these things are not done to the plants themselves, while pruning merely removes what is already dead, or what is not needed for the plant to thrive. To be sure, the tone in Question 11 is soft enough, but it has to be wondered if it is wholly consistent with Aquinas' position on human nature. Yearly correctly observes that "Aquinas cannot embrace a pure developmental model" (Yearly 1990, 79) and suggests that what is needed is a more rigorous application of rational reflection in the form of practical wisdom. However, Yearly does not say whether a teacher is required to help the learner apply rational reflection and exercise practical wisdom. In Question 11, Aquinas states clearly enough that a teacher can do so. Indeed, there is perhaps a hint of the firm hand of the teacher in his approval of St. Gregory's remark that the work of an active life includes "teaching the ignorant the word of wisdom." Indeed, given the fact that, as Yearly puts it, "inclinations to virtuous states co-exist with, and even connect to, powerful inclinations to bad states," a case can be made for a teacher to try to do the work of an angel, namely, to teach exteriorly but also interiorly in trying to weed out or at least subdue the 'inclinations to bad states.' If, as Aquinas maintains, the learner's intellectual light can be intensified by the angel's more perfect intellectual light, there is no reason why a teacher's more superior wisdom (on certain matters at least) cannot intensify the learner's, helping the latter to combat powerful inclinations to bad states. Perhaps this is what is meant by an *inspiring* teacher.

My readings of Aquinas and Mencius above, assuming that they are right, point to a number of issues that require further analysis. The comparison of Aquinas and Mencius indicates that Aquinas may be said to be splitting the difference between Mencius and Xunzi on the question of human nature and thus splitting the difference between the two on educational philosophy as well, although largely in favor of Mencius on both counts. However, by bringing in Xunzi, a question arises concerning the degree of firmness in moral education that is consistent with Aquinas' position. Related to this is the question of what adjustments need to be made to the developmental model that Mencius has in mind for the cultivation of the self to make it work for Aquinas. Should a teacher try to teach like an angel? As for Mencius, even if we accept that his developmental model was appropriate for his time, when there was a greater commitment than now to a set of social values, a question arises as to whether it remains appropriate in the modern context, when social cohesion is under considerable strain. Followers of Mencius might want to ponder whether a step towards Aquinas' implied position on moral education, which assigns a more extensive role to the teacher and makes teaching more a part of the teacher's active than contemplative life, is warranted. Of course, all bets are off for Mencius if it turns out that Aquinas is right in taking human nature to contain both good and bad elements, that some of the seeds in the human heart-mind would naturally sprout into noxious weeds, particularly in the current environment, which seems much more favorable for such seeds to sprout. If Aquinas is right about our inner constitution, and given "the present condition of the world," to borrow Heidegger's words, then while Mencius may not have to go as far as having to say, with Heidegger, that "only a God can save us" (Sheehan 1981, 45–46), he might have to go as far as saying "Only a sage-teacher can save us": saving us by being more interventionist in his or her teaching, intervening in the developmental process of the sprouts themselves.

End Notes

1. All page references given in text for this work are to Aquinas (1953).
2. "Question 11" does have a broader significance than moral education. I focus on moral education in order to facilitate my comparative reading.
3. See Gelfert (2006, 627–652). To be sure, this is in Aquinas as well.
4. Thus Aquinas writes: "For the moral character of some human actions is so evident that they can be assessed as good or bad in the light of these common first principles straightaway with a minimum of reflection. Others, however, need a great deal of consideration of all the various circumstances, of which not everyone is capable, but only those endowed with wisdom." Cited in Yearly (1990, 49) (ST I-II.100.1).
5. Translations of the *Mencius* and the *Analects* have been adapted from various sources, including James Legge, *The Works of Mencius* (New York, Dover Publications, 1970), and D.C. Lau, *The Analects* (London: Penguin Books, 1979).
6. This is not to deny that Aquinas has serious concern for social harmony. Indeed, it may be said that his natural law theory is deeply shaped by such concern.
7. See Knoblock (1994).

References

Aquinas. 1953. *Truth*, vol. II. Trans. James V. McGlynn. Chicago: Henry Regnery Co.
Donohue, John W. 1968. *St. Thomas Aquinas and education*. New York: Random House.
Gelfert, Axel. 2006. Kant on testimony. *British Journal for the History of Philosophy* 14(4): 627–652.

Heidegger, Martin. 1981. Only a god can save us. In *Heidegger: The man and the thinker*, ed. Thomas Sheehan. Chicago: Precedent Press.

Kant, Immanuel. 1963. Idea for a universal history from a cosmopolitan point of view. In *On history*. Trans. Lewis White Beck. Indianapolis: The Bobbs-Merrill Co.

Knoblock, John. 1994. *Xunzi: A translation and study of the complete works*. Stanford: Stanford University Press.

Lau, D.C. 1979. *The analects*. London: Penguin.

Legge, James. 1970. *The works of Mencius*. New York: Dover Publications.

Sheehan, T. (ed.). 1981. *Heidegger: The man and the thinker*. Chicago: Precedent Press.

Yearly, Lee H. 1990. *Mencius and Aquinas: Theories of virtue and conceptions of courage*. Albany: SUNY Press.

Exemplars for the Moral Education of Beginners in the Religious Life: Aquinas and Dōgen

Douglas K. Mikkelson

Interest in Thomas Aquinas *vis-a-vis* East Asian philosophical and religious thinkers has been building slowly but steadily over the last few decades. Evident parallels between the Aristotelian-Thomistic and Chinese Confucian traditions have encouraged efforts to compare Aquinas with figures such as Confucius and, especially, Mencius.[1] Indeed, in this volume Anh Tuan Nuyen demonstrates interesting parallels between Aquinas and Mencius on moral education. But if we narrow our focus on moral education somewhat and concentrate on the moral education of those 'leading the religious life,' East Asian parallels to that of Japanese Zen Buddhism, and especially towards the Zen Master Dōgen, present themselves as well.

We can begin with the observation that Aquinas (1221–1274) and Dōgen (1200–1253) were contemporaries who themselves pursued the religious life in their respective traditions. Catholic tradition labels persons living the monastic life in community and those living the mendicant life as friars as 'religious.' Sufficient similarities exist between the Catholic and Buddhist traditions for us to employ Catholic terminology such as 'religious,' 'the religious life,' 'monks,' 'nuns,' and so forth without undue violence to historical accuracy. Dōgen is an example of a Buddhist monk living in a monastic community; Aquinas was a mendicant whose communal life was centered in a community, though as a member of the Dominican order he also operated outside of it in the course of his activities.

Being religious (in the technical sense defined above), we should not be too surprised to discover that both men specifically addressed the moral education of those leading the religious life. Amidst the prodigious literary output of both men, we see considerable attention given to this topic. I will primarily limit my comparative discussion to two texts, namely *ST* II-II.186–189 of the *Summa Theologiae* and Dōgen's *Shōbōgenzō Zuimonki*,[2] though a few additions from outside these texts will enter the analysis as well, especially in these opening pages.

D.K. Mikkelson (✉)
Department of History, University of Hawai'i at Hilo, Hilo, HI, USA
e-mail: dougmikk@hawaii.edu

T.B. Mooney and M. Nowacki (eds.), *Aquinas, Education and the East*, Sophia Studies in Cross-cultural Philosophy of Traditions and Cultures 4,
DOI 10.1007/978-94-007-5261-0_7,

That this article intends, as the title indicates, to focus on the moral education of *beginners* in the religious life is the best reason for establishing these boundaries. Challenging as they may be to read today, both the *Summa Theologiae* and the *Shōbōgenzō Zuimonki* were intended as primers by their respective authors. As such they are logical places to look for the simple and introductory instructions offered to newcomers in their vocations.

"Having set out the general theory on vices, virtues, and other topics related to morals [i.e., in the *prima secundae* of the *Summa Theologiae*] we must turn now towards specific details about each." There are, Aquinas informs us, two ways to proceed:

> [One] is to look at the moral topics themselves, examining, for example, one or other particular virtue or vice; the second is to look at people in their respective callings, for example, to subjects and superiors, those pursuing the contemplative and those pursuing the active life, or other differences among people. Accordingly our own specific considerations will concern, first, themes related to all stations in life; secondly, details related to particular callings. (*ST* II-II, *Prologus*)

Accordingly, *ST* II-II.1–170 is concerned with "all stations in life," and *S.T.*II-II.171–189 with "particular callings." *ST* II-II.186–189 more narrowly addresses the calling of "religious," and thus provides a good focal point for our discussion.

As a pedagogical text, the entire *secundae* of the *Summa* can be seen as an introduction—albeit a very long one—on moral education. As the prologue of the *secunda secundae* makes clear, the education proceeds via an articulation of virtues and vices. Indeed, it would be plausible to explore the *secunda secundae* in relation to Dōgen's writings and conduct a comparative study of Aquinas and Dōgen on the virtues.[3] Certainly this discussion of virtues could be brought to bear specifically on the education of religious. But inasmuch as this essay intends to look more specifically at 'beginners' in the religious life, I propose that we may be able to discern a possible thread in both Aquinas and Dōgen that is interwoven within their thinking about virtues and vices, namely that of 'moral exemplars.'

That this is possible is suggested by a juxtaposed reading of the *Summa* and the *Shōbōgenzō Zuimonki*. Zen Buddhism is a tradition rife with instances of 'teaching by example,' and thus one is not surprised to see Dōgen repeatedly resorting to the use of moral exemplars in the course of his moral instruction. As for the *Summa*, let us begin with an introduction preliminary to the one found in the *secunda secundae*, namely the *prologus* to the *prima secundae*:

> Man is made to God's image, and since this implies, so Damascene tells us, that he is intelligent and free to judge and master of himself, so then, now we have agreed that God is the exemplar cause of things and that they issue from his power through his will, we go on to look at this image, that is to say, at man as the source of actions which are his own and fall under his responsibility and control.

Note how, prelimary to a discussion on morality, Aquinas reminds us (as spelled out previously in *ST* I.44.3) that God is the "exemplar" of all things. Since, as he also notes, humans are made in the divine image ultimately God is the exemplar of human beings. Placed prior to a discussion on morality, we readily see that God serves as the moral exemplar for human beings.

A precise and technical discussion of what Aquinas meant by exemplar, and how he employed it in his metaphysics, could be pursued,[4] but is unnecessary for our purpose here. The more general meaning we can get from the basic meaning of God as 'exemplar' here—that God is the ultimate model for moral human behavior—is sufficient to open up interpretive possibilities that offer points of comparison and contrast with Dōgen on the subject of the moral education of religious.

What model moral behavior of religious is supposed to look like is something that both Aquinas and Dōgen spell out in terms of virtues and vices, and can also be spelled out through categories such as precepts and rules and so forth. But we can also talk about moral exemplars and adopt an interpretive angle that goes to something more basic—namely, that Aquinas shows us many times what 'a good person' acts like in the course of providing a moral education. Especially for the beginner in the religious life, this would be a readily accessible thread of teaching.

A remark by Dōgen seems relevant:

> A man of old has said: 'Do not talk about the conduct of others if you don't resemble them.' This means that, without knowing or studying a person's virtues, one should not, upon seeing his weaknesses, conclude that he is a good person but suffers certain defects and does bad things. Look at just his virtues, not his shortcomings. This is the meaning of the saying: 'The gentleman sees the virtues but not the shortcomings of others.' (*SZ* 3.8)

In a manner parallel to what we see in Aquinas, in this passage Dōgen articulates what a good person is in terms of virtues and vices. Implicit in the passage, however, is that a good person is a model of behavior. That the good person is an exemplar is evident elsewhere in the text, wherein we are told to observe and associate with a good person, and in the course of doing so we will become good just as a man walking in the dew will get wet (*SZ* 4.14, 5.15). These passages exist alongside numerous others in the text that provide specific moral exemplars to emulate, and furthermore display normative standards for what constitutes good behavior.

At this point in the discussion we can envision three possible advantages to pursuing the matter of moral exemplars in a comparative study of Dōgen and Aquinas on moral education. First, Dōgen's focus on exemplars in the *Shōbōgenzō Zuimonki* can encourage us to push a bit further than we might otherwise do in interpreting Aquinas' moral pedagogy in the *Summa*, perhaps helping us to better appreciate how he communicates his teaching on the virtues. Second, we have the potential of opening up avenues wherein Aquinas can help us better understand how Dōgen uses exemplars. Finally, we have the possibility of identifying analogous moral exemplars between Aquinas and Dōgen.

Speaking of 'analogous moral exemplars' is useful here, because we are comparing two traditions that, for all their similarities, are also different in key respects. That they do not literally share moral exemplars is just the beginning point for understanding the need to pursue analogues. But it is also important to recognize that if we can discern significant similarities despite the differences, we might have justifiable reason to believe we have some good analogues we can use to bring to the task of a comparative study of Aquinas and Dōgen on the moral education of those leading the religious life.

We should take a moment to note how Aquinas and Dōgen proffer moral exemplars in their respective texts. A demonstration via Aquinas can emerge in a consideration of *ST* II-II.186.4 of the *Summa*, wherein the question is "whether perpetual continence (*continenta*)[5] is required for religious perfection."[6] The Second Opinion states: "[T]he first model of perfection for us (*primum perfectionis exemplar nobis*) is proposed via Abraham, to whom the Lord said, 'Walk with me and be perfect.' But the copy need not exceed the model. Therefore perpetual continence is not required for religious perfection." Aquinas, in the *sed contra* ('on the other hand'), begins by citing "what Saint Paul says" in II Corinthinas 7.1—"Let us cleanse ourselves from all defilements of the flesh and of spirit, perfecting sanctification in the fear of God"—leading towards the assertion that "therefore religious perfection requires continence." After his Answer, Aquinas' subsequent reply to the Second Opinion begins: "As Augustine says, 'The chastity of celibacy is better than the chastity of marriage, the second of which Abraham had practised, but he had the habit of both.'" This Reply then continues as a reinforcement of Aquinas' "Yes" answer to the question.

True to form, in *ST* II-II.186.4 Aquinas relies heavily on proof-texts to construct his argument, but how he employs them varies. Aquinas directly presents Abraham as a moral exemplar in the Second Opinion, based on a reading of Genesis 17.1. II Corinthians 7.1 appears in the *sed contra*, but note how this passage presents Paul's *teaching*, not Paul as a model of behavior. Finally, Aquinas uses a passage from Augustine's *De bono coniugali* ('The Good of Marriage') as a proof-text; this is also a teaching, but note how Augustine himself uses Abraham as an exemplar. This last example is an instance wherein Aquinas effectively advances a moral exemplar that is embedded within the words and teaching of Augustine.

As for Dōgen, he does not use exemplars embedded within teachings of others to any significant degree.[7] But like Aquinas, he also interweaves teachings and exemplars in order to advance his moral instruction. Furthermore the density of moral exemplars is greater in the *Shōbōgenzō Zuimonki* than the *Summa*.

Aquinas, we noted, viewed God as the exemplar of all things, and thus by extension the moral exemplar of all human beings. If the point of this essay is to suggest that Dōgen is particularly useful as an East Asian dialogical partner for a better understanding of Aquinas' instruction on the moral education of religious, we might ask if there is a plausible analogy one can offer to the Thomistic God here, and the none-too-surprising, simple answer is 'the Buddha' as moral exemplar. However, we need to recall a few basic points about Aquinas' conception of God and Dōgen's formulation of the Mahayana Buddhist understanding of the Buddha lest we be led in a false direction here. For sake of brevity and in order to keep this essay on track, I will seek to delineate only what appears to be essential in establishing and subsequently discussing these analogues.

Consonant with what he inherited from Catholic tradition, Aquinas adhered to a Trinitarian God, expressed as three *personae*: Father, Son, and Holy Spirit. This Trinitarian God is the exemplar of all creation. Furthermore, as all of creation proceeds from this Trinitarian God, so does Christ, the Second Person, proceed from the Father, while also returning to God in perfect unity. United with God, Christ is the perfect human and thus the exemplar for all humanity. (As expressed in *ST* III.50.1,

"Christ had a real body of the same nature as ours, a true rational soul, and together with these, perfect Deity.") As Thomas F. O'Meara puts it, "Jesus is both causal influence and model," (O'Meara 1995, 1254) and by extension we see how Jesus, both fully divine and fully human, is properly understood—in accord with what is maintained about God in the prologus of *ST* I-II—as the perfect moral exemplar for humanity.

Though undoubtedly a truncated account of Aquinas' conception of God (and perhaps also a somewhat unbalanced description via its emphasis on God and Christ as exemplar), hopefully this succinct account is sufficient for suggesting the proper analogue in Dōgen. Just as we needed to redirect talk about God to talk about the Trinitarian God, so we must redirect talk about the Buddha to the Triple Body of the Buddha. As a Japanese Buddhist, Dōgen inherited, and adhered to, the Mahayana interpretation of the *trikaya* (the Triple Body of the Buddha) which maintains that the Buddha is composed of the *Dharmakaya* (the body of reality), the *Sambhogakaya* (the body of delight) and the *Nirmanakaya* (the incarnated earthly body of the Buddha). This incarnated body is none other than Shakyamuni, 'the historical Buddha.'

Talk of the Shakyamuni as an incarnation of the Triple Body of the Buddha obviously calls to mind the Christian doctrine of the incarnation. Indeed, some Mahayana formulations stress the idea of the *Nirmanakaya* as proceeding from the *Dharmakaya*, an idea that seems to resemble the construction of Christ, the Second Person, proceeding from the Father in Aquinas. But Dōgen saw the *Nirmanakaya* as arising interdependently and simultaneously with the *Dharmakaya* and *Sambhogakaya*. Thus, unlike the Thomistic conception of the Trinitarian God in relation to Christ, the *Dharmakaya* does not "cause" the historical Buddha to be in the strict sense of the term; rather, all three *trikaya* realms arise interdependently within a web of causation.

If the difference in ontology here seems to move Dōgen's conception away from the historical Buddha as an appropriate analogue for the Thomistic Christ, it is more than made up for in the emphasis Dōgen places on Shakyamuni *vis-à-vis* other Mahayana (and Japanese) Buddhist traditions, which historically have tended to deemphasize him. As Hee-Jin Kim states, for Dōgen, "Shakyamuni Buddha is a historical person—an absolutely unadulterated, concrete human being, and the same historical person is the Buddha-dharma as well. [Thus,] *The historical Buddha became the prototype of the Buddha-dharma in Dōgen's thought*" (Kim 1975, 167, emphasis added). Given the multiple meanings of the term Buddha-dharma in Dōgen's thought, Kim packs more than one layer of meaning into the assertion that the historical Buddha "is Buddha-dharma as well," as we see in his subsequent analysis. But for our purposes, the most basic, fundamental meaning—that the historical Buddha is, for humanity, the exemplar for the realization of the truth the Buddha discovered—moves us significantly forward in our attempt to explore how exemplars serve to assist beginners within the religious life in their quest for moral education. One last link is necessary to establish Dōgen's Buddha as the proper analogue to the Thomistic Christ-as-moral-exemplar, namely the observation that for Dōgen, the realization of the Buddhist truth, expressed by Dōgen as the actualization of the enlightenment of the Buddha, entails not only realization of what truly is, but what truly is morally right.

Both Aquinas and Dōgen, therefore, posit a human as a moral exemplar for all other humans. These two figures, both in their being and becoming, are undeniably grounded and intertwined in an ultimate reality (the Trinitarian God for Aquinas, the Triple Body of the Buddha for Dōgen). As we saw with Christ, the Buddha also can be accurately described as both 'causal influence and model' for morality. For our purposes, we should also note that both Aquinas and Dōgen invoke their respective exemplars not only for humans in general, but in the specific instance of those leading the religious life. The *sed contra* of *ST* II-II.187.5 asserts that "It becomes religious to live in imitation of Christ"; *SZ* VI.12 informs us: "Zen monks are sons of Shakyamuni; they must learn the style of the Tathagata [lit. 'thus-come-one,' an epithet of the historical Buddha]."

Now we can turn to our selected texts and consider how Christ and the Buddha are employed as moral exemplars for religious in *ST* II-II.186–189 and the *Shōbōgenzō Zuimonki* respectively. We can begin with the following observation: just as Aquinas employs Christ both as a teacher of right morality and a moral human exemplar one should imitate, so does Dōgen employ the Buddha both as a teacher of right morality and a human exemplar one should imitate. Certainly the two *modi operandi* are closely related, yet if we look carefully the distinction gives us some purchase. In *SZ* 1.16, for example, Dōgen argues for the necessity of poverty for a monk. Near the middle of his presentation he quotes the Buddha as saying, "Possess nothing except your robes and bowl and give to starving people the leftovers from what you have begged." In this segment of the passage, Dōgen invokes the Buddha as a teacher of right morality. Yet a little further on in *SZ* 1.16 Dōgen seeks to bolster his argument further by presenting the Buddha as a moral exemplar:

> Did not even the Buddha offer twenty years of his life in this degenerate age! [...] Even though the Tathagatha [the Buddha] possessed supernatural powers of the greatest merit, he had to eat grain meant for horses to get through one rainy season. How can disciples in this degenerate age want things easier?

In this latter passage, the argument is advanced not via the teaching of the Buddha but by an account of the actual behavior of the Buddha in a particular moral situation: he exemplifies his commitment to poverty by resorting to eating grain meant for horses rather than abandoning his practice.

To be sure, the distinction between the Buddha as teacher of right morality versus the appeal to the Buddha's actions in the second passage as an exemplar for moral behavior is subtle here, and should not be overemphasized. Undoubtedly we are to understand the Buddha as someone who practices what he teaches, and vice versa. Yet in the progress of the *Shōbōgenzō Zuimonki*, Dōgen's moral pedagogy employs both modes in the articulation of a moral vision.

In the course of Aquinas' own argument in the *Summa* concerning the necessity of poverty for leading the religious life, we see something similar to what unfolded in the passage from *SZ* 1.16. The *responsio* of *ST* II-II.186.3 concludes with the assertion that for one who leads the religious life "the first foundation is religious poverty, whereby one lives without anything of one's own, as the Lord said, 'If thou wilt be perfect, go, sell all that you have, and give to the poor, and come...follow me.'" As Dōgen does with the Buddha in *SZ* 1.16, Aquinas employs Christ authoritatively via two modes in order to advance his moral instruction. Thus we have Christ-as-teacher (the content of the moral instruction following upon Aquinas' preferatory remark "as the Lord said"). We also have Christ-as-moral-exemplar, who is cited as exhorting others "to follow him," obviously not meant simply as the physical act of walking behind him on the same road but to follow his lived example on the same path of religious poverty.

On the subject of poverty for the religious, Dōgen employs the Buddha as a moral exemplar in his argument against the accumulation of riches:

> The Buddha was the son of a king and might have ascended the throne, had he so wished. Had he wanted, he could have bestowed treasures on his followers and furnished them with land. Why then did he give up his claim to the throne and become a beggar? It was to benefit people in the ages when the Law had declined and to encourage the practice of the Way that he set an example, by refusing to accumulate wealth and becoming a beggar himself. (*SZ* 3.7)

In Reply Four of *ST* II-II.186.3, Aquinas develops his own argument against riches, and again he draws from Christ-as-moral-exemplar as presented in Matthew 19.21, only this time he truncates the passage rather than elides it: "Wherefore our Lord said, 'Go, sell all that you have and give to the poor, and thou shalt have treasures in heaven." Aquinas also brings Christ-as-moral-teacher into the end of the argument: "So the Lord says [in Matthew 19.23] 'A rich man shall hardly enter into the kingdom of heaven.' This must be understood of one who actually possesses wealth, because of one who places his trust in riches...."

The above passage appears in the context of Aquinas arguing that a man surrounded by riches but nonetheless renouncing them to lead the religious life is blessed. A juxtaposed reading of these two passages from *SZ* 1.16 and *ST* II-II.186.3 provides potential matter for a comparative analysis of the material content of Aquinas and Dōgen on the accumulation of riches in relation to leading the religious life. We might, for example, observe that the Buddha, having been born into wealth and renounced it, appears to meet the criterion Aquinas establishes as a rich man who is blessed.

But we should also acknowledge that the establishment of appropriate texts for drawing out the best analogical material for exploring moral exemplars like Christ and the Buddha is not necessarily the best way to identify the best analogical material for comparative study on specific topics pertaining to Aquinas' and Dōgen's actual teachings regarding moral education. For example, in the passage from *ST* II-II.186.3 cited above, the first time Aquinas invokes the Matthean passage—in the conclusion of the *responsio* to the question—is offered in support of an argument seeking to show that, for the religious life, "the first foundation is poverty." If we focus on the best analogical material *for the content of Aquinas' statement*, rather than *the use of a moral exemplar to deliver it*, better analogical material can be found in *SZ* 3.11, which begins: "One day a monk came up and asked about the essentials of the Way. Dōgen instructed: 'Students must first of all be poor. If they have much wealth, they will certainly lose the desire for study.'"

Certainly, with the benefit of these texts, we could consider to what extent Dōgen could rightly be said to regard poverty as "the first foundation" for religious life. But it is a secondary consideration in this article, given that our real efforts here are to identify effective analogues demonstrating the similarities and differences Aquinas and Dōgen employ in the presentation of moral exemplars.

Like Dōgen, Aquinas not only regarded poverty as the first step necessary for the religious life, but maintained other requirements as well, such as continence; in the *responsio* of *ST* II-II.186.4 Aquinas asserts that "therefore perpetual continence is required for religious perfection, as is voluntary poverty." Yet in Reply One of the very same question he continues:

> Not only the perfection of poverty but also that of continence was introduced by Christ, who said [according to Matthew 19.12] 'There are eunuchs who have made themselves eunuchs for the kingdom of heaven'; and then added, 'He that can take it let him take it.' And lest the hope of attaining perfection be denied anyone, he admitted to the state of perfection even those he found joined in marriage. But husbands could not desert their wives without injustice, though men could renounce riches without injustice. Therefore when he met Peter, who was married, he did not separate him from his wife, although 'he kept John from marriage when he wished to wed.'

Aquinas seeks to answer the question of whether continence is required for the religious life, and the answer, albeit a qualified one, is yes. Similar to what we saw previously in *ST* II-II.186.3, Aquinas employs Christ-as-teacher by citing his words presented in Matthew 19.12, and then proceeds to invoke Christ-as-exemplar, noting that he admitted to the religious life even those who were married already.

Explicating exactly what Aquinas meant in his qualified Yes to the question—*prima facie* contradictory or paradoxical—would include reference to his point that right morality depends on various circumstances (see, for example, *ST* I-II.64.1, *ad* 2), a point on which Dōgen would wholeheartedly agree (see Kim 2004, 288). But this would require a detailed analysis beyond the parameters of this discussion. For our purposes, we can observe how Aquinas' answer includes Christ-as-exemplar adjudicating the same issue in the case of Simon Peter and John, son of Zebedee, two of his disciples and Apostles. In supporting his qualified Yes, Aquinas offers as authoritative not only Christ's judgements on Simon Peter and John, but the basic fact of Simon Peter and John having been admitted to the religious state. So we should pay close attention to how Aquinas constructed this passage. In effect, embedded within a presentation of Christ-as-exemplar is Peter-as-exemplar and John-as-exemplar.

Thus not only Christ, the perfect person, but such exalted yet imperfect personages like Peter and John can serve as human exemplars. This is consistent with the reasoning he employs in, among other places, *ST* I.44.3. Therein he discusses how, while God is the first exemplar of all things, yet "in things created one may be called the exemplar of another by reason of its likeness thereto." Thus, while Christ is the human moral exemplar par excellence, other human moral exemplars are possible precisely because, as indicated previously in the prologus to the *prima secundae*, "man is made in God's image."

To find the closest analogy in Dōgen to Aquinas' conjunction of Christ with two of his closest disciples as exemplars, we can begin by referring to another segment of *SZ* 3.7 cited earlier: "Monks should take care to follow the conduct of the Buddha and the Patriarchs. Do not covet wealth." Who precisely the Patriarchs are will be discussed momentarily. First we should note a point Dōgen makes in the course of exhorting his monks to vigorously pursue the religious life: "Even if you are not a person such as Subuti or Mahākāśyapa you should study the way in accordance with your capacity." (*SZ* 1.16) Just as Peter and John were two prominent disciples of Jesus, Mahākāśyapa and Subhuti were two prominent disciples of Shakyamuni. Mahākāśyapa was also a Patriarch of special importance for Dōgen, as evidenced by repeated references to him in Dōgen's massive *magnum opus*, the *Shōbōgenzō*. This is all the more evident when we know that Mahākāśyapa—according to Buddhist tradition—was the first to receive the Dharma transmission, the leadership of the Sangha (the Buddhist community) from the Buddha, as Dōgen notes in the "Bendowa" chapter of the *Shōbōgenzō*.[8] The obvious analogical figure here is Peter, who (according to the Catholic teaching tradition affirmed by Aquinas) was the first to receive the leadership of the Church from Christ.[9]

The functional analogy is significant inasmuch as Peter is the first in the line of apostolic succession, the body known collectively as the Popes, and Mahākāśyapa was the first in line of the Dharma transmission, known collectively as the Patriarchs. As Heinrich Domoulin notes, "Dōgen was convinced that the great way of the Buddhas and Patriarchs was not accessible without proper succession and inheritance" (Domoulin 1990, 63). That Dōgen held this view reflects some noticeable parallels he has with Aquinas. Aquinas sought to harmonize a bewildering inheritance, relying not only on scripture but the literature of classical antiquity. Dōgen himself was more 'classically-influenced' than the majority of his Ch'an (Zen) predecessors in China, and most probably all of his Zen contemporaries in Japan. As Ryusaku Tsunoda rightly notes,

> the Zen label he had little use for; it was the teaching of the Buddha, 'Buddhism,' with which he was drawn more and more to the Indian, especially the Hinayana sources, of this teaching than to the Chinese. (Tsunoda et al. 1958, 232)

Tsunoda is referring to Chinese Buddhism here, not Chinese traditions like classical Confucianism, a tradition Dōgen did draw upon, as we will discuss later. While it is true that Dōgen did indeed specifically include what are known as the Chinese Zen Patriarchs in employing moral exemplars, nonetheless he, like Aquinas, drew liberally from the early scriptures and history of his tradition to advance his moral instruction. Elsewhere in Dōgen's writings we read:

> Now [let me state the lineage of Zen Buddhism]. Our great teacher Shakyamuni transmitted the Law to Mahākāśyapa at a sermon assembly on Vulture Peak. It was correctly transmitted to the Venerable Bodhidharma through the successive patriarchs. He went over to China personally, and transmitted it to the Great teacher Hui-ko. This was the first transmission of the Law in China. Thus the Law naturally came to the Zen Master Hui-neng, the Sixth Patriarch Hui-neng. (*Bendowa*, 827)

Hui-neng is technically the 34th and last in the line of the Buddhist Patriarchs; Zen Buddhist tradition, however, usually refers to him as the Sixth [Zen] Patriarch, inasmuch as he was the sixth in line deriving from the figure regarded as the First Zen Patriarch, namely Bodhidharma (who was himself the 28th and last Indian Patriarch). Hui-neng is the Patriarch most often cited in Dōgen's writings; in *SZ*

3.14 Dōgen mentions him in the course of answering the following question: If parents are dependant on you, can you nonetheless leave them in order to become a monk? Using the example of Hui-neng, who originally supported his mother by selling firewood but eventually left her to pursue the religious life, ultimately Dōgen answers yes to this question. However, the passage also includes Dōgen's remark that if the man "could assure the comfort and livelihood of his mother," it would be a good thing; he also notes that Hui-neng "managed to obtain ten pieces of silver for his mother's food and clothing."

Hui-neng is the Patriarch most often cited by Dōgen in the *Shōbōgenzō Zuimonki*. The analagous figure we can locate in the *Summa* is Gregory the Great, the Pope most often cited by Aquinas in *ST* II-II.186–189; indeed, he is mentioned no less than 34 times within this text.[10] Notably, in *ST* II-II.189.6 Aquinas draws on Gregory in the course of answering essentially the same question Dōgen addressed in *SZ* 3.14; he constructs that question as "whether one should refrain from entering the religious life in order to support one's parents." Aquinas basic response, slightly different from Dōgen, is that children can enter the religious life except "in a case of necessity"—that is, in a situation where parents cannot be supported by someone other than their children. Yet, in contradistinction to Dōgen's employment of Hui-neng, in this passage, as elsewhere in the text, Aquinas is employing a passage from Gregory's writings as a proof-text; he never employs him as a moral exemplar.

As for Bodhidharma, in *SZ* IV.5 Dōgen informs us:

> The First Patriarch, Bodhidharma, came from the West and stayed at the Shao-lin Temple, awaiting for the opportunity to propogate the teaching. He sat gazing patiently at a wall, until Hui-k'o appeared in the last month of the year. Bodhidharma knew that here was a vessel of the Supreme Vehicle, guided him, and later transmitted both the robe and the teaching.

On first arriving in China, and prior to entering the Shao-lin monastery, Bodhidharma dwelled in a cave for 9 years. A similar event with another important exemplar appears in *ST* II-II.187.4 ("whether it is lawful for religious to live on alms."). Aquinas initiates his answer with the following *sed contra*: "Gregory says that the blessed Benedict, after leaving his home and parents, spent 3 years in a cave and lived on what was supplied for him by a Roman monk."

Gregory is never used as a moral exemplar, but in this instance Aquinas cites him in order to introduce another important moral exemplar in this text, namely Benedict of Nursia. Like Bodhidharma, Benedict's period of eremetic isolation in a cave was prior to his eventual entrance into a monastery. Benedict both created the best known Rule in Western monasticism as well as employing it in Monte Cassino, which he both founded and served as first abbot. But we are at a transition from a discussion of analogues between Popes and Patriarchs here, inasmuch as Benedict was an abbot but not a Pope, whereas Bodhidharma was a Patriarch but not an abbot. (Though we should note that the Dharma transmission proceeded from Bodhidharma to the Second Zen Patriarch Hui-ko, who himself became abbot of the Shao-lin Monastery.)

Of all the abbots present in *ST* II-II.186–189, Benedict is mentioned the most; of the five times he is mentioned, Aquinas uses him as an exemplar three times. Of all the abbots present in the *Shōbōgenzō Zuimonki*, Eisai (the abbot of Kenninji) is cited the most; in five references, Dōgen uses him as an exemplar three times.[11] A point of connection between these two figures can be found in *SZ* VI.15 and *ST* II-II.187.2. The former passage recounts an instance when Eisai receives a bolt of silk as a gift. He gives it to a Temple officer in order to buy some food. Just then, however, a request came from a certain layman. "I am embarrassed to ask, but I need two or three bolts of silk badly, and if you can spare any at all, I would deeply appreciate your letting me have some." The Abbot then took back the silk from the officer and gave it to the laymen.

In the case of someone in need, Dōgen approves Eisai's intervention in a matter of secular business, a point reinforced elsewhere in the *Shōbōgenzō Zuimonki*. That it is lawful for the religious to occupy themselves with secular business in instances of need is also supported by Aquinas, though the employment of Benedict as an exemplar in the *Summa* passage is a bit more complicated. We begin with the first Opinion presented in *ST* II-II.187.2: "It seems that it is not lawful for religious to be involved in secular matters, for Pope Boniface states in the aforementioned decree that 'Saint Benedict commanded them to be altogether free from secular affairs…'." True to form, Aquinas does not reject this passage, but contextualizes it; essentially, he sees Benedict's example as pertaining to secular business conducted from motives of avarice, whereas "from motives of charity, and with their superior's permission, they may occupy themselves with due moderation in the administration and direction of secular business." (*ST* II-II.187.2, *Responsio*). Furthermore, as in the *Shōbōgenzō Zuimonki* passage, in this line of argument Aquinas justifies the intervention of a religious in secular business as a matter of necessity. Thus Aquinas' reply to Opinion One—no doubt a reply he believes Benedict would approve of—reads: "Monks are forbidden to engage in secular matters out of avarice, but not out of charity."

So far, the exemplars we have explored have all involved what we can call people who are more or less 'religious' in both the broad sense we mean by it today and the more narrow technical sense—those leading the religious life. But both Aquinas and Dōgen also provide moral exemplars who are neither religious in the narrow sense nor primarily thought of as religious in the broader sense. Introducing these primarily secular individuals testifies to the broad cosmopolitanism of Aquinas' and Dōgen's moral pedagogy. 'Secular' figures not only contribute as moral exemplars throughout their literary output, but also to the specific instance of those in 'the religious state' as well.

Addressing the matter of whether poverty is required to lead the religious life (*ST* II-II.186.3), Aquinas poses Opinion Three as follows:

> 'Virtue consists in the mean,' as Aristotle says [in *Nichomachean Ethics* II.6]. But he who renounces everything by voluntary poverty does not seem to observe the just mean but rather goes to the extreme. Therefore he does not act virtuously, and hence this does not pertain to perfection of life.

Aristotle's virtue-ethics are critical to the composition of the *Summa*, and Aquinas liberally appropriates them in his construction of *ST* II-II.186–189. Aristotle is, of course, neither a Christian nor a 'religious' in the Thomistic sense, and even given that one can argue his thinking is religious at certain points, as when he presents an understanding of God as the Prime Mover, nonetheless he seems to be what we generally think of as a 'secular' philosopher.

In seeking an East Asian analogue to Aristotle, Confucius readily comes to mind. Comparative studies of Aristotle and Confucius on virtue-thinking can readily be found.[12] Confucius' thought is sometimes studied for content on subjects we commonly regard as religious, such as his attention to ritual (*li*).[13] But he is also generally understood as someone who is best described as a secular philosopher. Aquinas himself utilizes the distinction between the religious state versus the secular state, for example, in the opening lines of his *sed contra* to *ST* II-II.186.9:

> [T]he religious state (*status religionis*) is more secure than the secular state (*status saecularis*), and hence Gregory [in *Epistula ad Leandrum* 75, 511] compares secular life to a stormy sea, but religious life to a tranquil harbor.

Aquinas draws on Aristotle in order to address the matter of poverty for religious, and analogously Dōgen draws on Confucius in relation to the same issue: "According to Confucius, one should: 'In the morning hear the Way, in the evening die content.' Students today should emulate this attitude." (*SZ* II.16) Dōgen delivers this instruction as an introduction to the theme of this passage, namely a stress on the importance of "[not] planning for tomorrow's livelihood and hesitating to forsake what should be forsaken." Earlier in the text, Dōgen provided the very same quote (specifically identifying it as coming "from a non-Buddhist work") to support his argument for the necessity of poverty in a Zen monk (*SZ* I.16).

Aquinas' response to the countervailing opinion merits quotation in full:

> According to Aristotle, the just mean of a virtue is measured 'by right reason,' and not by the quantity of a thing. Therefore whatever can be done according to right reason does not become sinful by reason of quantity, but more virtuous. Now, it would be contrary to right reason if one were to use up all his goods through intemperance or for no useful purpose. But it is in accordance with right reason for one to renounce all his riches in order to give himself to the contemplation of wisdom. We read that even some of the famous philosophers did this. Thus Jerome says, 'The famous Crates of Thebes, once a very wealthy man, threw away a large amount of gold when he went to Athens to study philosophy, for he thought that he could not possess gold and virtue at the same time.' Therefore it is even more in accordance with right reason that a man renounce all his possessions to follow Christ perfectly. So Jerome says, 'Naked follow the naked Christ'. (*ST* II-II.186.3, Reply Three)

The reply culminates with an example of what we discussed previously: Christ as exemplar. However, neither in the reply above nor elsewhere in *ST* II-II.186–189 does Aquinas evoke Aristotle as an exemplar, as Dōgen used Confucius; rather, the words of his *Nichomachean Ethics* serve as a proof-text. Yet we clearly see the evocation of other philosophers as moral exemplars: "Even some of the philosophers did this"—i.e., engage in voluntary poverty. The conclusion of the reply also deserves attention. Aquinas evokes the exemplar of a secular philosopher engaging in voluntary poverty, and Aquinas stresses how much more it is "in accordance with right reason that a man renounce all possessions in order to more perfectly follow Christ"—in this context, Aquinas means to more perfectly pursue the religious life. This basic line of argument—namely, if voluntary poverty is morally advantageous for one in a secular state, how much more is it for one in a religious state—is also one advanced by Dōgen. Asked by a monk about the essentials of following the Way of the Buddha, Dōgen begins by asserting that "students must first of all be poor," and he illustrates his point with a story about a layman named Pang, who threw all of his possessions into the sea. He concludes his instruction thus: "Though a layman, he is known as a good man because he discarded his wealth in this way. How much more necessary is it then for a monk to discard his treasures!" (*SZ* 3.11.) Noteworthy here is how both Aquinas and Dōgen clinch this teaching for religious by employing moral exemplars who are in the secular state.

Elsewhere Dōgen observes that "even among laymen, the wise are aware of their responsibilities and perform their functions to the fullest. They seek no special rewards." (*SZ* 1.9) The remark follows upon an anecdote about a general named Lu Chung-lien, whom Dōgen offered as a moral exemplar. The general declined reward from his ruler for subduing enemies of the kingdom, declaring "A general's duty is to subdue the enemy; it is not to gain praise and possessions." Aquinas himself offers his own general as moral exemplar in the person of Judas Machabaeus "who fought with cheerfulness the battle of Israel, and got the people great honor"; he is an example of someone whose motivations were "not for worldly goods." (*ST* II-II.188.3, *Responsio*)

Perhaps one could postulate additional analogous secular moral exemplars, but I suspect at this point the law of diminishing returns applies. This may be true as well if we travel backwards into the 'religious' exemplars and try to establish more analogues between Christian abbots and Buddhist abbots and Popes and Patriarchs. Certainly there are abbots in the *Summa* other than Benedict and abbots other than Eisai in the *Shōbōgenzō Zuimonki*, but significant similarities among them do not seem as readily apparent. Nor are we likely to find Popes and Patriarchs as compellingly analogical as the disciples Peter and Mahākāśyapa.

Given the numerous references to other disciples of Christ in the *Summa*, analogues to the disciples of the Buddha might initially sound more promising. But beyond Mahākāśyapa, only three additional disciples of the Buddha are mentioned. Beyond mention of Subhuti in the same breath as Mahākāśyapa (see *SZ* I.16), nothing more is said about him; another disciple, Ksudrapanthaka, receives brief mention in a single sentence as an example of a dull person "who had trouble reading even a single line of verse" but nonetheless achieved enlightenment. But the third disciple is noteworthy:

> The decadence of monks is usually connected with wealth and rank. When the Buddha was alive, Devadatta's jealousy was aroused by the daily offering of five hundred cart-loads of provisions. He brought harm not only to himself but also to others. Why should true students of the Way wish to gain riches? (*SZ* 6.5)[14]

Devadatta, an evil disciple motivated by avarice to commit evil deeds, calls to mind Judas Iscariot. Chapter 12 of the Gospel of John recounts Mary of Bethany's anointing of Jesus with a costly perfume, and then relates: "But Judas Iscariot, one of his disciples (the one who was about to betray him), said 'Why was this perfume not sold for 300 denarii and the money given to the poor?'" (John 12.4–5) In *ST* II-II.188.7 Aquinas cites the subsequent text, John 12.6, which reads: "He said this not because he cared about the poor, but because he was a thief; he kept the common purse and used to steal what was put into it."[15]

Prima facie we seem to have the intriguing possibility of analogous 'anti-moral exemplars' here. Devadatta was a highly respected disciple who attempted to kill the Buddha; Judas was 1 of the 12 Apostles who committed actions resulting in the death of Jesus. Dante consigned Judas to the lowest circle of Hell, and according to Buddhist tradition Devadatta was condemned to long suffering in the hell realms. But now we should consider how Aquinas actually uses Judas:

> The Lord, who instituted poverty, taught this by his own example, for he had a 'purse,' entrusted to Judas, in which the offerings given him were kept, according to John [John 12.6—the passage quoted above]. Nor can one argue that Jerome says, 'If anyone objects, how is it that Judas carried money in his purse, we answer that Christ considered it unlawful to spend the money of the poor on himself,' that is, 'to pay the tax,' because his own disciples were first among the poor, and the money from Christ's purse was spent for their needs. Thus, John [4.8] says, 'His disciples were gone into the city to buy meats,' and that the disciples 'thought, because Judas had the purse, that Jesus has said to him: Buy those things which we have need of for the festival day, or that he should give something to the poor' (John 13.29). (*ST* II-II.188.7, *responsio*)

Unlike Devadatta in *SZ* VI.5, in *ST* II-II.188.7 Judas is not an anti-exemplar; rather, Aquinas' example of Judas in the above passage is part of his argument in defense of religious holding property in common. Thus whatever similarities we may draw between Devadatta and Judas generally, they do not function as analogous exemplars or anti-exemplars in this instance.

Yet the context wherein Devadatta and Judas appear deserves mention. Sustenance of the religious of the community, together with the necessity of poverty, are part of the larger backdrop to both of these passages. Even if the functional implementation of these two disciples are different, nonetheless the content of the instruction bears significant similarities.

If we push back further past the disciples, we arrive at the exemplars who launched the discussion, namely Christ and the Buddha. What this article provides is only an introduction to the analogous relationship between these exemplars. A cursory review of both *ST* II-II.186–189 and the *Shōbōgenzō Zuimonki* reveals numerous additional appeals to Christ and the Buddha as moral exemplars. Thus there is at least one set of analogues presented in this article that could be expanded upon. Closer attention may very well reveal in richer detail the employment of these analogous exemplars on similar subjects. Granted, following this approach, or seeking out other good analogues for moral exemplars in these texts, is not necessarily the best way to identify the best analogical material for comparative study on specific topics pertaining to Aquinas' and Dōgen's actual teachings regarding moral education, but it may help bring to the fore at least some of them.

Of course, one would not expect to find analogous exemplars utilized to the same degree or distributed across these texts in the same way. On the one hand, both

Aquinas and Dōgen give considerable space to issues regarding voluntary poverty, and accordingly we see analogous exemplars to a higher degree in these passages than elsewhere in the texts. On the other hand, Aquinas devotes a whole question to the matter of different kinds of religious life (*ST* II-II.188), and this is not a topic undertaken in the *Shōbōgenzō Zuimonki* to any significant degree.

Of course, when reading Aquinas and Dōgen we encounter many Popes and Patriarchs and abbots that do not lead to the discovery of significant analogies among them. Also, the *Summa* offers up exemplars from religious states and secular states that have no clear parallel in the *Shōbōgenzō Zuimonki*, and vice-versa. Perhaps with a closer look other possibilities could be proposed—for example, the Church Fathers *vis-a-vis* the famous old Zen Masters—but the differences between these two sorts of exemplars probably render the analogy too weak to be significant. Bishops for Aquinas and Emperors for Dōgen are examples of exemplars probably lacking plausible parallels in the opposing texts.

In conclusion, both Aquinas and Dōgen use exemplars either directly or through the teachings of others in order to provide a picture of what the 'good person' is. Attention to the broad array of moral exemplars may amplify awareness of the prodigious knowledge, cosmopolitan sophistication, and skillful means they bring to the task. When we consider the specific instance of exemplars advanced for the moral education of monks, we have reason to believe there are good analogical exemplars that facilitate the comparative study of Aquinas and Dōgen on the subject.

End Notes

1. Most notable is Lee Yearley's groundbreaking *Aquinas and Mencius: Theories of Virtue and Conceptions of Courage*. Yearley (1990).
2. Cf. Thomas Aquinas, *Summa Theologiae*. London: Blackfriars, 1964. *ST* II-II.186–189 refers to questions 186–189 of the *secunda secundae* of the *Summa*. I have altered the standard citation system by providing additions intended to indicate more precisely the position of a passage in a given question (see footnote six below). Furthermore, for the sake of accessibility I have followed the Blackfriars translation of the *Summa* as closely as possible; however, my own interpretation of the text has led me to make a few minor changes at certain points. The English text of the *Shōbōgenzō Zuimonki* (henceforth *SZ*) is taken from Masunaga (1971).
3. See, for example, Mikkelson (2005), 542–569.
4. For example, in *ST* I.44.3 Aquinas addresses the matter of "whether God is the exemplar cause of things" (*utrum Deus sit causa exemplaris rerum*) or whether there are other exemplars besides him. For a recent, detailed discussion of this and related topics see Doolan (2008).
5. "Thomas speaks of continence rather than the vow of chastity." Jordan Auman, OP, in Thomas Aquinas (1964, 115n).
6. Aquinas' *modus operandi* in the *Summa* is to list a series of points that *prima*

facie appear contrary to the answer Aquinas will give; English translations often identify these points as "Objections," although perhaps "Opinions" better captures his intent, since Aquinas does not always offer an unqualified "Yes" to the question, and usually seeks to harmonize his reply to them. Following upon the Opinions are a *sed contra* ("on the other hand"), an initial comment about the Opinions (which usually proposes a view at variance with the Opinions) and a *responsio* ("I answer that"), and finally a series of points that we can identify as "Replies" to each Opinion in turn.

7. Dōgen's embedded exemplars are typically unnamed. One exception appears in *SZ* V.7, wherein a priest named Hsueh-feng appears as a kind of anti-exemplar. A brief consideration of the question of anti-exemplars is discussed later in this essay.
8. The term Sangha can also have the more restricted sense of the community of Buddhist monks.
9. Analogous as Peter and Mahākāśyapa may appear in a functional sense, these exemplars Aquinas and Dōgen inherited respectively would seem to convey some fundamentally different messages in terms of the content of morality. Peter was criticized implicitly by Luke and explicitly by Paul for purported wishy-washy behavior and by members of the Jerusalem community for laxity in following moral dietary laws because of his decision to dine with gentiles; nonetheless, he was chosen by Jesus as his successor, the rock on which the Church would be built. Mahākāśyapa is generally regarded as receiving leadership of the Sangha because of his renown for ascetic self-discipline and moral strictness. Furthermore, in relation to the discussion above regarding Aquinas and Dōgens' respective views on the necessity of religious maintaining continence, Peter was married, but Mahākāśyapa was not.
10. How you count the citations of Gregory the Great depends partly on how you treat the repetition of a citation in a response that had first appeared in the corresponding Opinion to the question. Cf. Deferrari and Barry (1956), s.v. "Gregorius (the Great)," 144.
11. In the *Summa*: *ST* II-II.187.1; 187.2; 187.4; 189.5; 189.9 (moral exemplar in the first three); in the *Shōbōgenzō Zuimonki*: SZ.I.6; SZ 1.17; SZ II.2; SZ II.6; SZ V.6 (moral exemplar in the first, second, and fifth). SZ I.2 and II.5 mention Eisai in passing as a means of identifying another individual.
12. See, for example Yu (2007) and Sim (2007).
13. Confucius also inherited the legacy of *tian* (*tien*) (Heaven) from earlier Zhou Dynasty thought, which he regarded as the source of right morality. The religious character of his conception is well expressed in a passage from the Analects (II.4), wherein he informs us that "at fifty, I knew what were the biddings of Heaven, [and] at sixty, I heard them with a docile ear." See Waley (1938, 88).
14. Reihō Masunaga comments: "Jealousy over the 500 cart-loads of provisions, offered by a wealthy donor, motivated Devadatta to commit his evil acts" (Masunaga 1971, 118n).
15. The translation of John 12.4-6 provided here is from the New Revised Standard Version (1990).

References

Aquinas, Thomas. 1964. In *Summa Theologiae*, ed. Thomas Gilby and T.C. O'Brien. London: Blackfriars.
Deferrari, Roy J., and M. Inviolata Barry. 1956. *A complete index of the Summa Theologica of Thomas Aquinas*. Baltimore: The John D. Lucas Printing Company.
Domoulin, Heinrich. 1990. *Zen Buddhism: A history, volume two: Japan*. Trans. James W. Heisig and Paul Knitter. New York: MacMillan Publishing Company.
Doolan, Gregory T. 2008. *Aquinas on the divine ideas as exemplar causes*. Washington, DC: Catholic University of America Press.
Kim, Hee-Jin. 1975. *Dōgen Kigen: Mystical Realist*. Arizona: The University of Arizona Press.
Masunaga, Reihō. 1971. *A primer of Sōtō Zen: A translation of Dōgen's Shōbōgenzō Zuimonki*. Honolulu: The University of Hawaii Press.
Mikkelson, Douglas K. 2005. Aquinas and Dōgen and virtues. *Philosophy East and West* 55(4): 542–569.
O'Meara, Thomas F. 1995. Thomism. In *The Harper Collins encyclopedia of Catholicism*, ed. Richard O'Brien. San Francisco: HarperCollins Publishers Inc.
Sim, May. 2007. *Remastering morals with Aristotle and Confucius*. New York: Cambridge University Press.
Tsunoda, Ryoku, et al. 1958. *Sources of Japanese tradition*, vol. 1. New York: Columbia University Press.
Waley, Arthur (trans.). 1938. *The Analects of Confucius*. New York: The Macmillan Co.
Yearley, Lee. 1990. *Aquinas and Mencius: Theories of virtue and conceptions of courage*. Albany: State University of New York Press.
Yu, Jiyuan. 2007. *The ethics of Confucius and Aristotle: Mirrors of virtue*. New York: Routledge Publishing Co.

The Simplicity of the Ultimate: East and West

Joseph S. O'Leary

1 Introduction

To find an opening for dialogue, and for mutual critique, between Thomas and the great philosophies of India, it is best to focus on some fundamental or cardinal principle of his thought that presents a strong affinity with some equally central dimension of Indian thought. We need not look very far to find a candidate for this role. The third *quaestio* of the *Summa Theologica* (Aquinas 1964–1970 and Aquinas 1981, henceforth *ST*) is devoted to Divine Simplicity, a topic that had steadily risen to the top of the list of Thomas' concerns. Not only does this topic dominate the entire discussion of God in the subsequent 40 *quaestiones*, but it is grounded in the most central theme of Thomist metaphysics: the notion of being as act, and of divine being as *ipsum esse subsistens*, as the act of being subsisting in its pure and complete form. Divine simplicity "greatly affects the import of other things traditionally said of the divine nature" (Weigel 2008, 13). In Aquinas' hands it becomes the instrument for a revolutionary streamlining and 'simplification' of Christian doctrine.[1]

In the first half of this essay I shall reflect on the power and necessity of this teaching, and then I shall point out its problematic aspects: its logical difficulties, its apparent lack of phenomenological basis, its origin in metaphysics rather than Scripture, and the danger that it poses of simplifying God out of existence, as it were. In the second half of the essay I shall note how Buddhist and Vedantic thought was also held in thrall by the idea of ultimate simplicity, and how within Mahāyāna Buddhism an immense effort was made to reconnect *nirvāna* with the complexities of incarnate living. The fascination of simplicity and the logical and existential difficulties that it faced in both Europe and Asia constitute a heritage of thought that can be best received today by setting up a

J.S. O'Leary (✉)
Department of English Literature, Sophia University, 7-1 Kioicho, Chiyoda-ku, Tokyo 102-8554, Japan
e-mail: josephsoleary@hotmail.com

T.B. Mooney and M. Nowacki (eds.), *Aquinas, Education and the East*, Sophia Studies in Cross-cultural Philosophy of Traditions and Cultures 4,
DOI 10.1007/978-94-007-5261-0_8,

two-way philosophical interaction between the most lucid representatives of the two traditions, such as Nāgārjuna and Śankara, Plotinus and Aquinas.

2 The Fascination of Simplicity

All created beings are composite, since *what* they are is not identical to *that* they are, whereas the divine essence and the divine act of being are one and the same. Even in a purely spiritual creature such as an angel there is a composition of its essence and its existence, which are related as potency and act. However, *esse* is not a material to be poured into the formal frameworks provided by *essentia,* but is rather that which gives to *essentia* its actuality. An entity's essence can never be without its existence and vice versa; unlike the case of its matter and form (which are also related as potency and act).[2] The act of being is not beneath form but is a higher level of form; it is *maxime formale omnium,* "the most formal of all" (*ST* I.7. a.1), *formalissimum.* There is a certain simplicity in created being as well as in divine being, in that despite the immense diversity of created essences, all are actualized by being as act, which God has in its fullness. Thus the divine being is present and active at the heart of all created being, which radically depends on it. Yet Aquinas is careful to refute the pantheistic error of Amalric of Bena, who held that God is the *esse formale omnium,* being as the form of all things (*SCG* I, 26). We may suspect some subtle contradiction here; if God is the unlimited act of being, how can the being God confers on creatures be cleanly distinguished from his own being?

The idea of simplicity has often been rejected as logically untenable, and indeed the first major discussion of simplicity in Western philosophy, in Plato's *Parmenides,* underlines the logical antinomies it incurs. A determined reading, or misreading, of this discussion as sketching a constructive philosophy allowed the idea of simplicity, along with the idea of a positive infinity, to make its decisive breakthrough in Western philosophy in Plotinus' account of the One.[3] The vindication of simplicity imposes exigent logical claims, pushing thought to extremes of paradox. Its starkness breeds a fretfulness, as in Plotinus' attempt to give an inner life to the One by imagining it as *Causa Sui*: "he gives himself existence (*hupostêsas auton*)." (*Enneads* VI, 8, 16, Plotinus 1988) He already located the One as a hypostasis beyond being, but he now goes further, attempting under the veil of the *hoion* ('as it were') to think the impossible—"for it is impossible for a thing to make itself and bring itself into existence (*eis hypostasin agein*)." (*Enn.* VI, 8, 7) It is as if the One, despite its lack of substance (*ousia*), and despite its status as *dunamis pantôn*—the power of everything, on which everything depends for its being—were still too positively and blankly simply there, a heavy datum paralyzing the inquiring mind. When this hypostasis is imagined as giving rise to itself, it takes on a new vibrancy, and acquires a sharp profile as a dynamic self-grounding event.

Why and how did a self-sufficient, utterly simple reality bring into being the complex world we know? The answer is murky: it seems that a certain indeterminate intelligible matter—a no-thing—somehow comes forth from the One, and then gazes back at the One contemplatively, whereupon it crystallizes into a world of form, the Mind or *Nous.* (*Enn.* II, 4, 5) Iamblichus, dissatisfied with this, doubles the

One: "Prior to what is regarded as the first God and King (*to haplos hen*) is a God who is immovable and 'abiding in the solitude of His own unity.'" It is the lower One who is "father of Himself, self-begotten" and from whom derive entity and essence, the limit and the unlimited, and thence all things (Mathis 1992, 64–65).

In Proclus and Damascius, pure simplicity, the ultimate ineffable absolute, is pushed onto an ever higher shelf, while lower entities proliferate and germinate to bring about the different levels of cosmic reality. Hegel is close to Proclus when he writes that Plotinus

> did not advance from the unfathomable solitude of the One to its absolute freedom, in other words to the necessity whereby the Alone, who lacks nothing [...] nonetheless emerges from its solitude, advances and sacrifices itself in being other, thus affirming the absoluteness of its now personal freedom. (Cattin 2010, 161)

For Thomas the generation of the lower world does not demand any such elaborate negotiations between the One and the many. God is not beyond being, but is the fullness of being, pure act, and he brings finite beings into existence not by making them from a prior substrate but by creating them from nothing.

Gregory of Nyssa[4] and Augustine owe much to Plotinus in their convinced championship of divine simplicity, but Aquinas is his major heir, in that he thinks through the logic of the doctrine and grounds it in a vision of the simplicity of being. Plotinus' One was 'beyond being' and was known in an ecstatic 'touching' (*thigein*) that lies beyond conceptual knowledge. Aquinas, who understands God as the fullness of being, can marshal an elaborate conceptual discourse in which various aspects of the realm of being are predicated analogically of God. But Aquinas' thinking of being integrates many of the apophatic characteristics that marked Plotinus' tracking of the One and that were elaborated flamboyantly by Pseudo-Dionysius under the influence of later Neoplatonism. Those repelled by the starkness of the claim that "*Deus est purus actus non habens aliquid de potentialitate,* God is pure act having nothing of potency" (*ST* I.3.2), may find some comfort in the fact that Aquinas' discourse on the divine simplicity proceeds under the sign of the negative. "Because we cannot know of God what he is, but what he is not, we cannot consider the manner in which God is, but rather the manner in which he is not" (*ST* I.3, prologue). Questions 3–11 of the *Prima Pars* offer then not a positive account of divine attributes but a denial of any attributes that would imply lack of simplicity, perfection, goodness, unity in God or the limitations of temporality, location or change. Is there a buried contradiction between the simplicity claimed for God and the great complexity of what is nonetheless said about him, or between the emphasis on proceeding by negations and removals and the quite positive content of the analogical statements on God's goodness, intelligence, knowledge and will? If there is, Thomas has succeeded in keeping it out of sight.

3 Religious Resistance to Simplicity

Divine simplicity is not a biblical doctrine but a metaphysical refinement, formulated as a matter of faith at the Fourth Lateran Council in 1215, which taught that God is "a completely simple substance or nature" (*substantia seu natura simplex omnino*).[5]

Scripture indeed provides ample textual basis for the doctrines of divine unity, immutability and perfection, but neither the Councils nor St Thomas come up with any direct scriptural attestation for the notion of divine simplicity. Unsurprisingly, biblically grounded theologians deplore the way that in scholasticism simplicity was "exalted to the all-controlling principle, the idol…devouring everything concrete" (Barth, 1957, 329).[6] Many theologians and philosophers today are in outright revolt against the doctrine of divine simplicity, not so much for logical as for existential reasons. They want a God who can suffer and be affected by his creatures, a God who knows potency and possibility—a 'God Who May Be'[7]—, or a God of process, functioning as a cosmic serendipity, risking failure if the creation fails.[8] In classical theology all these dimensions are looked after by the incarnational economy, in which God assumes human suffering through the hypostatic union of the divine Word with the human Jesus, while any suggestion that the divine nature itself changes or suffers is kept at bay. Hyperpersonalistic, indeed tritheistic, accounts of the Trinity thrive on forgetting what Aquinas never forgets: that all processions, persons, relations, notions within the divine must be constituted without compromising the utter simplicity of God.

The basic defence of Thomist Trinitarian language is to say that since God is pure act there can be no division or composition in his nature.[9] The act of God's being, of God's self-knowing, of God's self-loving is one and the same. The one single divine act of being is Father, Son and Spirit depending on whether it is viewed as origin, as self-knowing or as self-loving. That this pure act can differentiate itself by two processions, according to intellect and according to will, with no prejudice to its utter simplicity, testifies to the extreme richness of this reality, just as the multiplicity of the divine attributes does. Just as God's goodness, power and wisdom are each identical with the divine essence, so Father, Son and Spirit are each identical with the divine essence, but with the new twist that they are distinct modes of self-relation of that essence.

The closest analogy to divine simplicity is the simplicity of the human mind, which is one single conscious entity though embracing a constantly shifting repertory of objects of awareness. God knows all things by knowing himself: "*Alia autem a se videt non in ipsis sed in seipso, inquantum essentia sua continet similitudinem aliorum ab ipso,* God sees things other than himself not in themselves but in himself, insofar as his essence contains the likeness of things other than himself." (ST I.14.5 *resp.*) In Plotinus, the self-thinking Mind, despite its high degree of integration, is not purely simple, and is far beneath the One. Aquinas carries off the *tour de force* of identifying the pure simplicity of God with the perfection of His intellect.

A rarefied and airless construction, it might be thought, offering no foothold for the religious imagination. And yet the notion of divine simplicity does elicit religious feelings, of a peculiar quality, which may be remote from the feelings Scripture creates. The contemplative and imaginative fascination of divine simplicity is given full-bodied expression at the climax of the most sublime poem of Western culture, where Dante images God as a minute point radiating "a light so keen that the eye on which it blazes needs must close."[10] A point is the simplest imaginable reality in the physical order, or rather it is so simple that it escapes from that order and remains an ideal geometrical construction or postulate. Talk of divine simplicity has always risked reducing God to an abstraction, or even proving God to be impossible. The image of divinity as potent concentration in a single point, immobile but omnipotent, without extension but omnipresent, exerts a hypnotic fascination. But to some it may seem excessively remote and even sinister, a black hole absorbing all life into itself. The cult of divine simplicity may seem a sublime fetishism, alien to the more incarnate and dynamic sense of divine presence in Scripture.

The Thomist idea of the utter simplicity of the subsistent act of being itself came to be felt as a screen against the simple reality of the divine, so that Eckhart (Eckhart 1984, 1986–1987, 1995) devotes his thought to the godhead in its inner essence 'before' it is differentiated into Trinity. Indeed, the godhead is close to the Plotinian One, preceding the God who is being.[11] This idea of the pure godhead is something like the nirvanic aspect of the divine, where all the differentiations on which Trinitarian doctrine concentrates fade away into ineffable mystery.[12] Talk of God in terms of being or in terms of Trinitarian distinctions is talk of God still turned toward worldly comprehension, and it does not have purchase on the unfathomable mystery of the Godhead in itself.[13]

Eckhart's leap beyond names and forms of God to the undifferentiated ocean of divinity, the holy nothingness of the 'godhead,' which is stripped of all positive characterizations, might seem an option for pure abstraction. But he can be read in a phenomenological sense, as renouncing names and forms in order to remain close to immediate phenomena. In the context of a life centered on the *imitatio Christi*, his stripping away of theological language could be the expression of practical realism comparable to that of the Zen masters. His quest for the naked essence of the godhead could be a hyperbolic demonstration of impatience with abstractions that impede access to the living God here and now. His apophatic language would then be that of a *Lebemeister* rather than a *Lesemeister*, a master of living rather than a master of reading. His active reflection, expressed in gestures, removals, strippings, renunciations, involves his very existence. The 'ebullition' of divine being is reflected in the energy with which he combines contemplation and action. His negation of truth and knowledge, the Good and will, being and the name 'God' is sustained by a purposeful striving that controls the energy of his life as much as of his speech.[14] Like Zen in Buddhist tradition, Eckhart effects a phenomenological turn in Neoplatonic tradition, bringing it to bear on common things. Eckhart becomes a phenomenological thinker of the intimate *esse* of things, as he "transforms the natural theology of St. Thomas into mysticism" (Lossky 1998, 31).

Restlessness with the seamless Thomistic vision of being and of God can also be felt in Nicholas of Cusa, who strives to empty the Thomist absolute of its positivism as an affirmation of being, by thinking back to an instance higher than being, which would be the condition of possibility of being and of God as being. For Cusanus this is the *posse,* the 'can' that constitutes the inner dynamic core of being, including the being of God. Both Eckhart and Cusanus can be seen as bringing the Thomist conceptions home to a prior dimension, marked by a still greater simplicity and peacefulness, but also by a more dynamic capacity for engagement with common things. The doctrine of creation itself is no longer merely a sudden communication of being to finite essences but becomes an expression of the inner life of the divinity.

The Western heritage of metaphysical thought about the divine simplicity is a problematic heritage. We feel it to be a precious heritage, and we sense that its inner tensions are full of instruction. Yet at the same time it can be felt as a cumbersome white elephant, far estranged from the living questions of contemporary thought. Would it double our problems if we turned to Indian thought, where we find an equally lofty and equally tension-ridden tradition of reflection on simplicity? Or can comparison of the two traditions create an illuminating perspective, in which the possibility of their retrieval and overcoming in a contemporary reinterpretation opens up? If so, we could find a new sense of simplicity, not as a recondite conundrum, but as a quality lodged at the heart of our real-life experience.

4 Bringing *Nirvāna* Down to Earth

Indian thought has often known the fascination of an ultimate reality characterized by utter simplicity: "All schools of Indian philosophy recognize the existence of something absolute that exists without conditions and is eternal or timeless. For the Sāmkhya, this is *purusa*, for the Vedānta it is *ātman/brahman*, and for the Jains it is *adharma.* For the Buddhists, *nirvāna* falls into this category" (Walser 2004, 193).[15] Of these, the Buddhist and Vedāntic absolutes (Lamotte 1962 and 1976, Saddhatissa 1985, Thibaut 1962) offer the closest affinities with the simplicity of the One or of God in Western metaphysics.

In early Buddhism ultimate reality is characterized in starkly negative terms, as in the *Udāna*: "There is, o monks, a not born (*ajāta*), not come into existence (*abhūta*), not made (*akata*), not composed (*asamkhata*)" (Woodward 1985, 98).[16] *Nirvāna* is one of those dharma that are *asamskrta*—not (*a*) made (*krta*) by reuniting (*sam*) parts or conditions, thus 'non-confected,' 'incomposite,' or 'unconditioned.' If some Western philosophers begin from a simple absolute, Buddhism starts more empirically from the given phenomenon, always composite, and argues back to what is not composite.[17] In the Theravāda school *nirvāna* is the only such incomposite, free of the qualities of arising, changing, and perishing, and putting a final end to the poisons of greed, aversion, and delusion. Other schools count dependent co-arising (the basic ontological texture of worldly reality) and its synonym *tathatā* or thusness as an incomposite, but the Theravādins object that this is a norm, not an entity, and moreover is associated with the realm of *samsāra*. The Sarvāstavādins propose three incomposites: space and the two *nirvāna* (with remainder and without remainder). The candidate for absolute simplicity that garners the widest consensus in early Buddhism is *nirvāna* without remainder, the condition attained by the Buddha at his death. Unsurprisingly, little can be said about this. If the human mind can offer an analogy of divine simplicity for St Thomas, in Buddhism the self is a composite of the five *skandhas* or 'heaps' of *dharmas*, all of which are ever-changing dynamic processes, and all of which vanish when one enters *nirvāna*: "There is no measure to him who has gone to rest; he keeps nothing that could be named. When all dharmas are abolished, all paths of speech are also abolished."[18] Among the 14 metaphysical questions that the Buddha refused to answer were the following: Does the *Tathāgata* (the nirvanized Buddha) exist after death, or not, or both, or neither? Nāgārjuna teaches that since all things are empty of substantial existence, they are ultimately as ineffable as *nirvāna*, so that one cannot ultimately say of anything that it is empty, non-empty, both, or neither. (*Madhyamaka-kārikās* 22, 11, see also Bodhi Bhikkhu 1995 and 2000, Walshe 1987, Bugault 2002, Conze 1970, and Garfield 1975)

The great revolution in Buddhist thought brought about by the Mahāyāna movement affects the way *nirvāna* is envisaged. It remains a condition of ultimate simplicity and peace: "Impermanent surely are conditioned things. It is their nature to rise and fall. For, having been produced, they are stopped. Their pacification brings ease."[19] In the Madhyamaka philosophy, all claims to being or substantiality (*svabhāva*) are dismantled—through logical confutation and meditative insight—in order to let the reality of emptiness (*śūnyatā*) emerge, in a nirvānic "quiescence of fabrications" (Nāgārjuna, *Madhyamaka-kārikās* 25, 24), free of the antinomies that beset all our categories and free of any clinging to delusive notions of substantial existence and identity. To realize emptiness and to know ultimate truth (*paramārtha-satya*) is to see through the discriminations and objectifications of the conventional, illusory world (*samvrti-satya*), in an insight (*prajñā*) that lies beyond speech and conceptuality. For a Buddha no objects exist; even the compassion of a Buddha is an objectless compassion.[20]

But the reality envisaged is no longer a stable absolute instance. Mahāyāna excludes from the realm of ontological possibility such a figure as the God of Aquinas or even the self-sufficient One of Plotinus. Or at least it challenges us to subject these figures to an 'emptying' that might give them a different mode of presence, building on the immanentist omnipresence of both. Free from all finite determinations, these figures of ultimate reality are present not as massive substances, but as subverting any finite category or image; they are present at once everywhere and nowhere, like *nirvāna*. *Nirvāna* itself cannot be absolutized. *Prajñā* perceives that "the dharmas are neither existent nor non-existent, neither arising nor passing away"; this is "the absolute purity that destroys all consideration of dharmas" (Lamotte 1944–1980, 2199). The contrast between incomposite *nirvāna* and composite dharmas becomes inoperative, since both composites and incomposites are empty of inherent existence. On the one hand, Mahāyāna attributes to all dharmas the quality of non-arising, which makes all dharmas *asamskrta* (in their ultimate reality). On the other hand, to attribute to the unconditioned any character at all, be it non-arising, is to make it a conditioned dharma; so the *asamskrta* become themselves *samskrta.* But this is a contradiction, forcing us to drop this way of thinking altogether, and to recognize that the dharmas are neither composite nor incomposite, conditioned nor unconditioned, absolute nor contingent. Their only character is their absence of character, an unobjectifiable emptiness, which it is a mistake to set up as an unconditioned, incomposite entity such as the *nirvāna* idolized in earlier Buddhism. Sharing the same emptiness, all dharmas are equal (*samatā*), and there is no difference to be drawn between *samsāra* and *nirvāna.* There is no utterly simple Absolute to be talked about.

Thomist negative theology could well integrate the techniques of Madhyamaka while allowing them to etch the form of the biblical God. The contradiction between

Thomist theism and Buddhist atheism begins to disappear when we realize that "śūnyatā represents the form of all absolutism" (Murti 1960, 237). Indeed, within Indian tradition a reappropriation of Madhyamaka for purposes similar to those of Aquinas can be studied in Śankara, who had imbibed the Madhyamaka methods and outlook through his teacher's teacher Gaudapāda and who provides a model of how to use and overcome Madhyamaka negativity.[21] Since the ultimate in Vedānta is characterized as being (*sat*) and consciousness (*cit*), it offers a substantial bridge for comparison with Aquinas. Later Vedāntic thinkers made the radical negation of Madhyamaka rhyme with the pleroma of undifferentiated being[22]; a similar dynamic can be found in Pseudo-Dionysius.

Śankara's *sat* is not an abstract vision of the essence of Being, but means concrete existing.[23] To overcome the illusory differentiations of our world is to discover the vitality of unrestricted being, and the unity between one's own deepest being, *ātman,* and the deepest being of all, *Brahman*, which is incomposite, entirely simple, omnipresent.

> Being is simple, with an absolute simplicity (*ekarūpa,* literally, uniform; *ekarasa,* having only one flavor, one essence; *sanmātra,* mere being, being pure and simple). Being is full (*pūrna*), with an absolute plenitude and density. Being is infinite (*ananta*). But no matter what aspect of it one takes, it is first and foremost tranquil, in repose. (Lacombe 1937, 34)

Methods deriving from Madhyamaka now serve to dismantle delusive 'superimpositions' in order to let this fullness of being emerge. This is not the erasure but the liberation of finite being. Each finite being is seen as "overflowing its limit in the very moment that this limit is posited, spreading itself freely near and far, melting into the others" (Lacombe 1937, 37).[24]

To many, this was a more attractive doctrine than the sheer emptiness that Nāgārjuna leaves us with, and it won the day in India as the fortunes of Buddhism declined. Indian commentators often find that the resemblances between Mahāyāna and Vedānta are so great as to make them almost indistinguishable: "The difference between māyā and Brahman is the difference between percepts bifurcated into subject and object and those same percepts experienced nondualistically. This is why Mahāyāna could equate samsāra and nirvāna" (Loy 1988, 68).[25] Mahāyāna reveals all beings as sharing the same nirvanized mode of being, so that we are already living *nirv na* here and now, and need only awaken to the fact.

Wherever we locate the difference between Buddhism and Vedānta, the nonduality common to both seems at first sight radically incompatible with the sharp distinction in Aquinas between the infinite and simple God and His finite, composite creatures. Does a stark dualism then confront an equally thorough monism? Perhaps a basis can be found for a spiritual dialogue between the two pictures if we consider that the divine simplicity is what allows the divine to be infinite and omnipresent, and to embrace all things, much as a Buddha's supreme and simple knowledge can course through all conventional realities. As Eckhart distinguished the essential godhead from the God of Trinity and Creation, Haribhadra (eighth century) distinguished between two dimensions of the *dharmakāya*, the ultimate reality of Buddhahood: the *svabhāvika-kāya* in which it exists in and for itself, and the *jñāna-kāya* constituted by the Buddha's knowledge of 21 sets of pure dharmas. In the latter, the Buddha is grasped as turned toward the world—though not in the sense of involvement in it as in his enjoyment-body (*sambhoga-kāya*) and apparitional body (*nirmāna-kāya*). Could we not suggest that the entire realm of metaphysical differentiations is traversed by a single reality having only one flavor, one essence, which we can call being or consciousness or liberation or emptiness? The discovery of this one thing in every experience does not abolish the diversity of the real but brings out its deepest sense in each particular experience.

What mystical writers suggest is that opening up to the presence of the divine, far from bringing an ever tighter grip on God's identity and on one's own, entails relinquishing the securities based on conceptual mastery in order to enter into the immediacy of life, wherein it may seem equally valid to say with the Vedantists: 'Atman is Brahman' and to say with the Buddhists: 'No Atman, No Brahman.' Here the divine simplicity is not a metaphysical conundrum but a phenomenological evidence. Metaphysical definitions of God are a hedge of negative prescriptions set about this encounter. Necessary as they are, they do not close in on the phenomenality of God as grace. In Zen Buddhism, simplicity becomes a matter of living here and now, casting off all intellectual discriminations that impede this. Thus Ma-tsu (709–788) can say "This very mind is Buddha," and on another day, "No mind, no Buddha," and on yet another day, "There is nothing at all." The subtle arguments of the different traditions lead us to a place where they become superfluous, since the simple reality has become manifest.

5 Conclusion

Can Thomas and Eckhart be seen as leading to the same place as Nāgārjuna and Śankara? Not immediately, perhaps; and a debate set up between these traditions is likely to be endless, rather than resolving itself into a simple chord. Great minds of East and West have been spellbound, and frustrated, by the postulate—or experience?—of an ultimate simplicity. An intelligent retrieval or critique of either tradition today is impossible without intensive consultation of the other.

The idea of a completely simple supreme reality has imposed itself not only on speculative thinkers in the lineages of India and Greece but also on mystics and contemplatives in both monotheistic and Oriental religious traditions. It remains confined, no doubt, to cultures marked by high philosophy, and even in those cultures it has lost much of its authority and self-evidence. Should we shake off this idea as a metaphysical imposition that prevents us from living life to the full in all its glorious complexity? Or does it contain some precious core of insight that makes its study of more than historical interest? In both East and West simplicity becomes intimately associated with what is nearest to hand—with the faithful presence of gracious ultimacy here and now. This offers the promise of an existential and phenomenological regrounding of the idea, an approach to it 'from below.' Metaphysical cogitations about ultimacy may become self-canceling as they lead us back to where we are. Once we have arrived there safely, we can let the cogitations go, as skillful means or rafts that have served their purpose. Such is the stance of the Zen master, and of the Christian who identifies with the incarnate Christ in the conditions of everyday life. Yet the rethinking of simplicity must also engage the logical and dialectical rigor of thinkers such as Thomas, Eckhart, Nāgārjuna and Śankara on its own terms. The speculative argument continues, but with a greater awareness of the historical and cultural pluralism of the languages of simplicity, and a greater willingness to overcome them when they fall prey to frozen abstraction.

End Notes

1. In the *Summa Contra Gentiles* (Aquinas 1955–1957, also translated in Aquinas 1975, henceforth *SCG*), Aquinas begins his account of God with the attributes of immutability (I, 13) and eternity (I, 15), in Augustinian style. The features discussed in the ensuing chapters (I, 16–27) are grouped in a tighter synthesis under the heading of simplicity in the *ST*; see Weigel (2008, 30–34).
2. See the luminous discussion of essence and existence in Wippel (2007, 123–51).
3. See Halfwassen (1992).
4. Discussed in O'Leary (2007, 307–37).
5. This was in reaction to the Albigensians, whose Manichean dualism undermined divine simplicity. The doctrine that God is *simplex omnino et incommutabilis substantia spiritualis* is repeated in Vatican I's Dogmatic Constitution on Catholic Faith (ch. 1) in 1870, probably with reference to speculative Catholic theologians such as Anton Gunther (1783–1863; condemned in 1857), who had tried to introduce a dynamic pluralism into the idea of God, under the influence of Hegel and Schelling; see O'Meara (1982).
6. Barth is quoted by Radde-Gallwitz (2009, 19).
7. See Kearney (2001); for a critique see O'Leary (2006, 185–207).
8. See Gordon D. Kaufman, criticized in O'Leary (1996, 180–185).
9. I thank Claude Geffré for this point.
10. "*un punto vidi che raggiava lume/acuto sì, che'l viso ch'elli affoca/chiuder conviensi per lo forte acume*" (*Par.* 28, ll. 16–18); Dante (1982, 313).
11. The Plotinian aspect is strongly emphasized in Pasqua (2006, 33–76).
12. "The formal emanation in the divine Persons is a type of 'boiling,' and thus the three Persons are simply and absolutely one"; Meister Eckhart, *Lateinische Werke* III, p. 291, quoted in McGinn (2001, 75).
13. "Some of his sermons, at least, challenge the ultimacy of the Christian Trinity by inviting the believer into the 'God beyond God,' that is, 'into the simple ground, into the silent desert, into which distinction never gazed, not the Father, nor the Son, nor the Holy Spirit'" (McGinn 2001, 79, quoting Sermon 48 in *Deutsche Werke*).
14. "*Tollitur omne velamen...sicut etiam velamen boni, sub quo accipit voluntas, velamen veri, cum quo accipit intellectus, et universaliter velamen ipsius esse* (Every veil is removed...including that of the Good, under which the will receives,

and that of the True, with which the intellect receives, and universally that of Being itself)"; *Lateinische Werke*, IV, p. 114, quoted in Lossky (1998, 193).
15. For *adharma,* see Glasenapp (1964, 153–54). For *purusa* in Sāmkhya, see Larson and Bhattacharya (1987, 73–83). For Advaita Vedānta, see Biardeau (1969, 39–55).
16. For the following discussion, see Lamotte (1980, 2182–186); Kapani (2005, especially 131–143).
17. See Bareau (1951, 4).
18. *Sutta-nipāta* 1076, quoted in Conze (1970, 79).
19. *Mahāpārinirvānasūtra*, as quoted by Conze (1970, 73). Conze calls this verse "the very epitome of Buddhist thought, and worthy of prolonged reflection."
20. See the discussion of this idea in Viévard (2002, 239–45).
21. Śankara created an original system, which may owe as much to Madhyamaka Buddhism as to the Upanishadic texts he treats; see Angot (2007, 13–40), for the argument that the continuity between the Upanishads and Śankara is largely a retrospective illusion. The system of advaita (non-dual) Vedānta should not be imagined to pre-exist, even in latent form, in the Upanishads.
22. I thank François Chenet for this information.
23. See Angot (2007, 132–38).
24. However, "the simplicity of being in itself is not to be sought at the point where the essences, saturated with their own being, are identified one with the other through their superabundance, but must be located at a far earlier point, when what for us is a plurality of essences renounces beforehand, as it were, the benefits of that plurality." (Lacombe 1937, 37).
25. Loy draws on Zen to show how this nondualistic thinking is cashed in human experience and action.

Glossary

Advaita Vedānta The philosophy based on the Upanishads, perfected by Śankara in the seventh century CE. Central tenet: the non-duality of *ātman* and *Brahman* (self and absolute).

Analogy Knowledge of divine attributes on the basis of the finite likeness of creatures to their infinite divine cause.

Apophatic theology Negative theology, which proceeds by negation (*apophasis*) and removal (*aphairesis*).

Dependent origination The basic ontology of Buddhism, the arising of phenomena in mutual dependence.

Dharma A phenomenon or element; supreme reality; the teaching of the Buddha.

***Dharmakāya*, dharma-body** The ultimate reality of a Buddha, as the embodiment of supreme reality; contrasted with the other two Buddha-bodies.

Emptiness, *Śūnyatā* The lack of self-existence in all dharmas.

Hypostatic Union The teaching of the Council of Chalcedon, 451 CE, that the human and divine natures of Christ are united in one hypostasis.

Madhyamaka or Mādhyamika The central philosophy of Buddhism, formulated by Nāgārjuna in the second or third century CE, vindicating the emptiness of all dharmas as taught in the Perfection of Wisdom sūtras.

Mahāyāna The 'great vehicle' Buddhism expounded in a series of sūtras in the early centuries CE.

Nirvāna Release from the painful condition of worldly existence and the cycle of rebirths.

Samsāra Worldly existence in its pain, impermanence, and bondage.

Skillful means, *Upāya* Teaching devices of a Buddha, accommodated to the capacity of the hearers.

***Svabhāva,* self-existence** The illusion of permanent substantial identity.

Trinity The Christian doctrine, finalized in 381 CE, that the one God subsists in three hypostases.

Zen (Chan) Monastic, meditative Buddhism developed in China, which foreswears dependence on linguistic and conceptual mediation, aiming at direct insight.

References

Alighieri, Dante. 1982. *The divine comedy: Paradiso*. Trans. Charles S. Singleton. Princeton: Princeton University Press.

Angot, Michel. 2007. *Taittirīya-Upanisad avec le commentaire de Samkara*. Paris: Collège de France.

Aquinas, St Thomas. 1955–1957. *Contra Gentiles: On the truth of the Catholic faith,* ed. Joseph Kenny. New York: Hanover House.

Aquinas, St Thomas. 1964–1970. *Summa Theologiae*. Trans. Thomas Gilby et al. New York: McGraw-Hill.

Aquinas, St Thomas. 1975. *Summa Contra Gentiles*. Trans. Anton C. Pegis et al. Notre Dame/London: University of Notre Dame Press.

Aquinas, St Thomas. 1981. *Summa Theologica, complete English edition in five volumes*. Trans. Fathers of the English Dominican Province. Allen Christian Classics.

Bareau, André. 1951. *L'Absolu en Philosophie bouddhique: Évolution de la notion d'asamskrta*. Paris: Tournier et Constans.

Barth, Karl. 1957. *Church Dogmatics II/1*. New York: Scribner's.

Biardeau, Madeleine. 1969. *La philosophie de Mandana Miśra*. Paris: Maisonneuve.

Bodhi, Bhikkhu (trans.). 1995. *The middle length discourses of the Buddha*. Boston: Wisdom.

Bodhi, Bhikkhu (trans.). 2000. *The connected discourses of the Buddha*. Boston: Wisdom.

Bugault, Guy (trans.). 2002. *Nāgārjuna: Stances du milieu par excellence*. Paris: Gallimard.

Cattin, Emmanuel. 2010. *Vers la simplicité: Phénoménologie hégélienne*. Paris: Vrin.

Conze, Edward. 1970. *Buddhist thought in India*. Ann Arbor: University of Michigan Press.

Conze, Edward (trans.). 1983. *Perfect wisdom: The shorter Prajñāpāramitā Texts*. Totnes, Devon: Buddhist Publishing Group.

Garfield, Jay (trans.). 1995. *The fundamental wisdom of the middle way: Nāgārjuna's* Mūlamadhyamakakārika. Oxford: Oxford University Press.

Halfwassen, Jens. 1992. *Der Aufstieg zum Einen: Untersuchungen zu Platon und Plotin*. Stuttgart: Teubner.

Kapani, Lakshmi. 2005. La dialectique du dépassement et le dépassement de la dialectique en Inde. In *Catégories de langue et catégories de pensée en Inde et en Occident*, ed. François Chenet, 125–170. Paris: l'Harmattan.

Kearney, Richard. 2001. *The god who may be: A hermeneutics of religion*. Bloomington: Indiana University Press.

Lacombe, Olivier. 1937. *L'Absolu selon le Védânta*. Paris: Geuthner.

Lamotte, Étienne (trans.). 1944–1980. *Le traité de la grande vertu de sagesse de Nāgārjuna*, 5 vols. Louvain-la-neuve: Institut Orientaliste.

Lamotte, Étienne (trans.). 1962. *L'Enseignement de Vimalakīrti*. Louvain-la-neuve: Institut Orientaliste.

Lamotte, Étienne, and Sara Boin-Webb. 1976. *The teaching of Vimalakīrti*. London: Pali Text Society.

Larson, Gerald James, and Ram Shankar Bhattacharya. 1987. *Encyclopedia of Indian philosophies IV: Sāmkhya, a dualist tradition in Indian philosophy*. Delhi: Motilal Banarsidass.

Lossky, Vladimir. 1998. *Théologie négative et connaissance de Dieu chez Maître Eckhart*. Paris: Vrin.

Loy, David. 1988. *Nonduality: A study in comparative philosophy*. New Haven /London: Yale University Press.

Mathis II, C.K. 1992. Parallel structures in the metaphysics of Iamblichus and Ibn Gabirol. In *Neoplatonism and Jewish thought*, ed. Lenn E. Goodman, 61–76. Albany: State University of New York Press.

McGinn, Bernard. 2001. *The mystical thought of Meister Eckhart*. New York: Crossroad.

Meister Eckhart. 1984ff. *L'œuvre latine de Maître Eckhart*, ed. Alain de Libera. Paris: Éditions du Cerf.

Meister Eckhart. 1986–1987 (repr.). *Die deutsche und lateinische Werke*, ed. Ernst Benz, Josef Quint et al. Stuttgart: Kohlhammer.

Meister Eckhart. 1995. *Selected writings*. Trans. Oliver Davies. London: Penguin.

Murti, T.R.V. 1960. *The central philosophy of Buddhism*. London: Allen & Unwin.

O'Leary, J.S. 1996. *Religious pluralism and Christian truth*. Edinburgh: Edinburgh University Press.

O'Leary, J.S. 2006. Questions to and from a tradition in disarray. In *After god: Richard Kearney and the religious turn in continental philosophy*, ed. John Manoussakis, 185–207. New York: Fordham University Press.

O'Leary, J.S. 2007. Divine simplicity and the plurality of attributes (*CE* II 359–386; 445–560). In *Gregory of Nyssa: Contra Eunomium II: An English version with supporting studies*, ed. Lenka Karfíková et al., 307–337. Leiden: Brill.

O'Meara, Thomas F. 1982. *Romantic idealism and Roman Catholicism: Schelling and the theologians*. South Bend: University of Notre Dame Press.

Pasqua, Hervé. 2006. *Maître Eckhart: Le procès de l'Un*. Paris: Éditions du Cerf.

Plotinus. 1988. *Enneads*. Trans. A.H. Armstrong, vol. VII. Cambridge, MA: Harvard University Press.

Radde-Gallwitz, Andrew. 2009. *Basil of Caesarea, Gregory of Nyssa, and the transformation of divine simplicity*. Oxford: Oxford University Press.

Saddhatissa, Hammalawa (trans.). 1985. *The Sutta Nipāta*. New York: Humanities Press.

Thibaut, George (trans.). 1962. *The Vedānta Sutras of Bādaryana with the commentary by Śankara*. New York: Dover.

Viévard, Ludovic. 2002. *Vacuité (śūnyatā) et compassion (karunā) dans le bouddhisme madhyamaka*. Paris: Collège de France.

von Glasenapp, Helmuth. 1964. *Der Jainismus*. Hildesheim: Olms.

Walser, Joseph. 2004. *Nāgārjuna in context*. New York: Columbia University Press.

Walshe, Maurice (trans.). 1987. *The long discourses of the Buddha*. Boston: Wisdom.

Weigel, Peter. 2008. *Aquinas on simplicity: An investigation into the foundations of his philosophical theology*. Oxford: Peter Lang.

Wippel, John F. 2007. *Metaphysical themes in Thomas Aquinas II*. Washington, DC: The Catholic University of America Press.

Woodward, F.L. (trans.). 1985. *Minor Anthologies*, vol. II. London: Pali Text Society.

Aquinas and Locke on Empiricism, Epistemology, and Education

Cecilia Wee

1 Introduction

Aquinas and Locke have a number of points in common with respect to their theory of knowledge. For one, both are well-known for their emphasis on the senses as a foundation for knowing. For another, there are significant commonalities between Aquinas' account of *scientia* (often translated as 'knowledge')[1] and Locke's account of knowledge. Both accounts hold that this ultimate form of ratiocinative human knowing[2] is to be distinguished from 'lesser' epistemic states, such as belief and opinion.

Given their shared stance on the origin of human knowledge, and their similar views on the nature of such knowledge, would their theories of *learning* be similar? Answering this question is the task of this paper. In doing so, I hope in part to contribute to the history of ideas by connecting Locke's views on knowledge and learning to those of Aquinas. More importantly, I hope to examine the ways in which the espousal of a particular theory of knowledge can shape its concomitant theory of learning, and more particularly, the role of the teacher in such learning.

The paper will have the following structure. In Sect. 2, I give an account of Aquinas' understanding of *scientia*, and show how it shapes his theory of learning and the role of the teacher in learning. In Sect. 3, I do the same for Locke's account of knowledge. I then argue that there is a small, but ultimately crucial, difference between Aquinas' and Locke's accounts of learning which concerns the role of the teacher. I argue that this is traceable to a difference in their theories of knowledge. Section 4 concludes and briefly relates my findings to the role of the teacher in present-day multicultural, multivalent contexts.

C. Wee (✉)
Department of Philosophy, National University of Singapore,
3 Arts Link, 117570 Singapore, Singapore
e-mail: philimtn@nus.edu.sg

T.B. Mooney and M. Nowacki (eds.), *Aquinas, Education and the East*,
Sophia Studies in Cross-cultural Philosophy of Traditions and Cultures 4,
DOI 10.1007/978-94-007-5261-0_9,

Before I begin, let me enter a few caveats regarding the aims and scope of this paper. First, note that I shall mean by 'theory of knowledge' a philosopher's particular views of the origin, scope, and nature of knowledge. With regard to the 'theory of learning', I shall be concerned primarily with the question of what the *process* of learning involves, and in particular, with the conditions under which one may be said to have learnt something new. Second, the present paper will focus primarily on Aquinas' and Locke's views on *natural* knowledge (i.e., what we would now encompass under knowledge of the natural sciences). The paper will not deal with their moral, political or theological views except insofar as they impinge on the former. Finally, given the considerable body of exegetical work on both Aquinas and Locke, there are, as one might expect, significantly differing views as to the precise content of their philosophical positions. It is well beyond the scope of this paper to enter into these discussions. Instead, I adopt what are fairly standard, and textually not unreasonable, readings of Aquinas and Locke. Where necessary, I will offer textual or other evidence to support my readings.

2 Aquinas on *Scientia* and the Teacher

Aquinas is commonly seen as accepting the Scholastic dictum, thought to derive from Aristotle, that 'there is nothing in the intellect that is not first in the senses.'[3] He clearly holds that knowing is ultimately founded on what is given to us by the senses. An understanding of Aquinas' theory of knowledge and knowledge acquisition cannot be accomplished without acknowledging the strong influence of Aristotle on his views. Like Aristotle, Aquinas was a metaphysical realist who held that the material world was composed of various objects with real natures. Again, like Aristotle, he thinks humans cannot grasp these real natures without sensory contact with the relevant objects.

For Aquinas, the route by which a human comes to grasp a real nature is as follows. Suppose that Mary has sensory perception of Rover. Mary's sensitive soul acquires the form of Rover, and this form, for Aquinas, is the sensible species of Rover. On the basis of this sensible species, Mary's intellectual soul then acquires the intelligible species of a dog. It is usually taken that this acquisition of the intelligible species involves a process of abstraction from the sensible species, whereby Mary eliminates the non-essential features of, say, specific doggy smell, brown fur, pointed ears, and by this procedure comes to grasp the real nature of a dog.[4] In doing so, Mary would be said, in more contemporary parlance, to grasp the real nature of all dogs insofar as dogs form a *natural kind,* that is, a kind found in the natural world.[5] (Other such natural kinds would include cats, gold and water.)[6]

For Aquinas, a key goal of the pursuit of natural knowledge is to attain understanding of the various real natures possessed by objects that belong to the various natural kinds found in the physical universe, and then to use this understanding to formulate various propositions putatively aimed at describing features of this universe. According to Aquinas, the highest form of ratiocinative knowing is *scientia*. Aquinas holds that, when one has a complete and certain grasp of something, one has *scientia* with respect to that thing, and others of its kind. The criteria for *scientia* are stringent. *Scientia* includes only (1) whatever is *per se nota*—that is, known through itself or self-evident, and (2) whatever can be derived through chains of reasoning from that which is *per se nota*. That is, *scientia* thus encompasses within its domain that which is fundamental or indemonstrable, and that which is demonstrated on the basis of indemonstrable truths.

Macdonald points out that Aquinas' account of *scientia* has to be understood within the context of his metaphysical realism and his view that the universe is composed of things possessing real natures waiting to be discovered. For Aquinas, "the structure of demonstration…is isomorphic with the metaphysical structure of reality: immediate, indemonstrable propositions express metaphysically immediate facts, whereas mediate, demonstrable propositions express metaphysically mediate facts"; and again, "the facts expressed by immediate propositions are such that once one conceives their terms, one is aware of the proposition's truth" (Macdonald 1993, 170). For example, assume that the proposition 'Dogs are animals' is an indemonstrable truth. Aquinas would hold that inspection of the concepts of 'dog' and 'animal' (derived from sense-experience) would make it evident that this statement is necessarily true. This necessary truth in turn rests on facts that obtain about the real natures, respectively, of dogs and animals. Thus, if Mary truly cognizes the real nature of a dog, and again of an animal, she would cognize a dog to be an animal.

Commentators on Aquinas' theory of knowledge tend to focus on his account of *scientia*. However, it should be noted that while *scientia* is certainly the highest kind of epistemic state attainable by humans, Aquinas had room in his account for lesser epistemic states. For example, he allows that certain kinds of reasoning (e.g., those whose conclusions are based on premises not known to be certain or those using non-deductive argument forms) can provide us with probable conclusions. Such probable, but not certain, conclusions bring about beliefs or opinions, but not *scientia*.[7]

Aquinas' accounts of the role of the teacher and the nature of learning are, in turn, very much grounded in his metaphysical realist position, and his account of how we obtain knowledge of real natures through sensory apprehension. Gilson notes that Aquinas' tract on teaching, the *De Magistro*, is notable at first sight for its conclusion that "no one teaches any one else anything" (Gilson 1948, 9). In the *De Magistro*, Aquinas goes through various arguments which suggest that there is no role for the human teacher in the learning process. Some of the difficulty in finding any significant role for the teacher is rooted in Aquinas' account of *how* one comes to know. For Aquinas, the original human route to *scientia* begins with interaction with the natural environment. It is through gaining the sensible species of a natural object through sense-perception, and grasping the intelligible species of that object in the intellect, that the knower comes to comprehend a real nature. On this basis, one then forms universal propositions, which when clearly and completely cognized, come to be included under *scientia*.

For Aquinas, the knower really knows—that is, is in possession of *scientia*—when she has fully comprehended, and clearly cognized, not only the real natures involved, but the universal propositions that she forms on their basis, and any other propositions established to be true through subsequent chains of reasoning. *Scientia* thus cannot be attained by blindly receiving and reproducing a series of claims. Rather, *scientia* is only attained when one has worked through, internalized, and fully understood through one's own God-endowed reason, each and every step by which one comes to assert a particular claim.

Thus, at first sight, there seems to be no room for a teacher—for every particular truth that the knower knows, the knower must come to know and recognize that truth to be so *in and for herself*. As Gilson notes: "It may be the same truth, but everyone who knows it, knows it through his own intellect, so that, ultimately, he is his own teacher" (Gilson 1948, 8–9).

As mentioned, the *De Magistro* accords considerable space to examining how the human teacher *does not* play a role in the learning accomplished by the learner. Nevertheless, after some prolonged discussion, Aquinas allows that there is a role for the human teacher after all. The teacher, he argues, can instruct by eliciting the intelligible species in the student's soul, without the latter having first acquired the sensible species from sensory interaction. He writes:

> In the pupil, the intelligible forms of which knowledge [i.e. *scientia*] received through teaching is constituted are caused directly by the … intellect and mediately by the one who teaches. For the teacher sets before the pupil signs of intelligible things, and from these [are derived] the intelligible likenesses [that] exist in the passive intellect. Hence, the words of the teacher, heard or seen in writing, have the same efficacy in causing knowledge as things which are outside the soul. For from both the [intellect] receives intelligible likenesses, although the words of the teacher are more proximately disposed to cause *scientia* than the things outside the soul, in so far as they are signs of intelligible forms. (*DM* 11:1)

For Aquinas, then, the teacher helps shorten the process involved in the acquisition of *scientia*. She does this by drawing out, via words which act as signs, the intelligible species into the knower. Thus, the knower bypasses the initial step of acquiring the sensible species in her sensitive soul, as well as the next step of abstraction from this sensible species by the intellect. Instead, the intelligible species is caused in her mediately by means of the teacher's words.

But Aquinas emphasizes that the student knower does not have *scientia* if she merely parrots the teacher's claims without understanding. She can attain *scientia* only if she genuinely understands and internalizes the content that her teacher imparts, and this involves going through the same processes of reasoning that the teacher herself goes through:

> Whenever anyone is taught by another, the learner must examine the concepts of the teacher, so that in this way the pupil's mind may reach *scientia* through the same reasoning process which the teacher's mind uses. (DM 11:3, 4)

So if she internally goes through the same processes of reason that the teacher does, she can be said to have learnt from the teacher.

One significant feature of Aquinas' account of knowledge and its acquisition needs to be highlighted. Aquinas has often been called an epistemological optimist,[8] insofar as his account of the acquisition of *scientia* does not concern itself much with skeptical arguments, of the kind later put forward by Rene Descartes,[9] against the possibility of such acquisition.[10] There is, at least *prima facie*, a stability and in-principle accessibility to the natural universe of Aquinas in which learning takes place—this universe is one of real natures, and one can, *en principe*, acquire knowledge of these real natures either through processes of reasoning based on the starting point of either sensory contact or the teacher's instruction.

This is not to say that no account at all is taken by Aquinas of the potential difficulties that the seeker after truth might encounter in his search. Aquinas accepts that we do not easily attain *scientia*. For instance, certain kinds of reasoning such as inductive reasoning cannot provide us with *scientia* but with the lesser states of belief or opinion. Again, Aquinas holds that, even when we follow the procedures prescribed for the acquisition of *scientia*, we will only succeed in the full acquisition of the truth in 'paradigmatic' cases[11] (by which he presumably means those cases in which all the relevant procedures have been flawlessly conducted). Indeed, the difficulty of attaining *scientia* is emphasized in the Prologue to the Apostle's Creed, where Aquinas notes, rather bleakly, that "our manner of knowing is so weak that no philosopher could perfectly investigate the nature of even one little fly."[12]

Nevertheless, that Aquinas gives an account of *scientia*, and clearly assumes in the *De Magistro* that both teacher and student *can* eventually attain *scientia*, is an indication that he thinks *scientia* may be difficult, but not impossible, to attain. What Aquinas does not explore is the question of whether conditions might obtain that might make it *utterly impossible* for the learner (or teacher) to ever attain *scientia*. For example, he does not ask whether there are aspects of the natural world that are opaque or inaccessible to the human learner, so that she can never have *scientia* about them. Again, he does not consider the possibility that our senses are so unreliable that they cannot at all serve as the foundation by which we attain *scientia*.

For Aquinas, truths about the natural world are, at least in principle, always available to the seeker of such truths. This can be seen in his characterization of the role of the teacher: "Whoever teaches another leads him to the truth and so causes truth in his soul." (DM 11:3, 6) Aquinas assumes here that the teacher is one who causes truth in the soul of her student: he does not question that one can indeed ultimately attain the truth, either through the teacher's signs—or indeed by means of one's senses.

In the next section, I will look at the views of Locke, who also held similar views on the structure of knowledge and of knowledge acquisition, and indeed on the nature of learning. I try to explain why, despite their similar views on these issues, they had divergent views on the role of the teacher.

3 Locke on Knowledge and Education

Aquinas is one of the most significant figures (if not the most significant figure) in medieval philosophy. Locke was a major figure of the Enlightenment, often seen as a key figure of modernity. It may therefore be surprising, at first sight, how much the two have in common with respect to their theories of knowledge and their views on education.

For Locke, the highest form of knowledge is obtained through intuition. Whatever is intuited is self-evident, and is immediately known to be true. Locke's notion of intuition thus closely parallels Aquinas' consideration of indemonstrable (*per se nota*) truths. Again, Locke distinguishes between intuition (which is self-evident) and demonstration (which is the derivation of a further conclusion on the basis of self-evident truths), in much the same way that Aquinas distinguishes between indemonstrable truths and the demonstrable truths that are based on them. Locke also accepts that intuition and demonstration are the highest epistemic states that the seeker after truth can achieve (though he does hold that demonstration is a notch lower than intuition in terms of certainty). Similarly, Locke also admits, as Aquinas does, of other epistemic states such as belief or assent, where the thinker is not certain that the proposition holds, but rather sees it as merely probable that it does.

In sum, then, for Locke, truths derived from intuition and demonstration are included under knowledge in much the same way that Aquinas includes demonstrable and indemonstrable truths under *scientia*. For both, the epistemic states of knowledge/*scientia* are seen as significantly different from states such as belief or assent. As Wolterstorff notes, for Locke, "knowledge is an act or state of mind fundamentally different from belief or assent. Knowledge…is perceiving some fact. Believing is *taking* or *presuming* some proposition to be true" (Wolterstorff 1996, 46). Thus, Thomistic *scientia* and Lockean knowledge involve recognition of *what actually obtains*, whereas belief or assent are only what one thinks likely to obtain.

Again, Locke shares with Aquinas the view that the edifice of knowledge/*scientia* is built upon what is received by the senses:

> Let us then suppose the mind to be, as we say, white paper, void of all characters … How comes it to be furnished? … Whence has it all the materials of reason and knowledge? To this I answer, in one word, from EXPERIENCE. In that all our knowledge is founded; and from that it ultimately derives itself. (Locke 1975, 104)

Lockean knowledge is founded upon *ideas of sensation* (which derive from sensory contact with the universe) or *ideas of reflection* (which involve introspection of one's mental states). This corresponds roughly with Thomistic *scientia*, which is founded upon sensible forms in the sensitive soul (arising from sensory contact with the universe). While Lockean ideas of sensation may not be entirely isomorphic with Thomistic sensible forms or sensible species, both Locke and Aquinas accept that knowledge/*scientia* are ultimately based on what we gain from the senses.

Again, at the next stage, where Aquinas is usually held to accept that the seeker after *scientia* acquires intelligible species or intelligible forms by abstraction from sensible species, Locke will hold that the seeker after knowledge acquires abstract ideas by a process of abstraction from sensory input. Further, recall that, in the case of self-evident or indemonstrable truths such as 'Dogs are animals', Aquinas holds that it is inspection of the concepts of 'dog' and 'animal' that make it evident that this universal claim is true. For Locke too, knowledge is "the perception of the connexion and agreement, or disagreement and repugnancy, of our ideas" (Aaron 1955, 238). Thus, we intuit (as Locke would say) that 'All dogs are animals' insofar as we can clearly see the agreement between the idea of a dog and the idea of an animal.

There is thus altogether a great deal of similarity in Locke's and Aquinas' theories of knowledge, of what it is that one can be said to know and how one comes to know that. What then of Locke's view of education—does Locke's theory of learning also bear significant similarities to Aquinas' theory of learning?

Locke is well-known for his crucial contributions to the theory of education, and in particular on the education of children. Influenced in his views on this issue by Montaigne and Rabelais, he was in turn himself an influence on figures such as Rousseau, whose *Emile* bears the imprint of his educational views. Though his views on education may seem somewhat conservative today, his recommendations were very radical in his own time, and they helped shape educational theory as it is today. This paper will not be concerned with the bulk of his educational views, but only with that portion that pertains to his account of the process by which one may be said to *learn*, and to attain knowledge.

Cranston notes that, for Locke, a sound education crucially involves that one develop the ability to "judge and discriminate between good and bad arguments". Cranston quotes Locke:

> When man by use hath got this faculty of observing and judging of the reasoning and coherence of what he reads, and how it proves what it pretends to teach; he is then and not till then in the right way of improving his understanding. (Cranston 1957, 244)

Locke thus holds that the process of reasoning is important for the process of learning. We do not *really* learn until we have brought our reason to bear on what is being taught or propounded.

Tarcov notes that Locke's *Conduct of the Human Understanding* can be seen as a self-help manual for adult education. In it, Locke details what one should do to avoid succumbing to "prejudice, unexamined first principles, reliance on authority, belief in infallibility, partiality, passion, interest," and so on (Tarcov 2003, 84). In *Conduct*, Locke stresses that one must not accept on trust whatever is delivered to one, but subject it to the rigorous inspection of reason. Locke would thus be opposed to blind and 'rote' learning. Instead, we should use and strengthen our natural reason, so that we will know *for ourselves* whether a particular claim or proposition is to be accepted, or to be rejected because of various deficiencies.

Locke's account of learning and knowledge acquisition thus bears some resemblance to that of Aquinas. Aquinas accepts that the learner, whether she is taught by the teacher or engaged in self-learning, must engage in the processes of reason to attain *scientia*. Without such active engagement, she does not attain the latter. The use of one's own reason is central to learning for both philosophers. Thus, despite the divide of several centuries, and the perhaps more important distinction that one was a medieval philosopher, and the other a philosopher of the Enlightenment, Locke and Aquinas did share substantial similarities in their views on knowledge, knowledge acquisition, and the nature of learning. However, there is one crucial difference in their views on learning, which relates to the role of the teacher in learning.

We can characterize this difference best by examining the prescribed *attitude* for the learner towards the teacher in their respective views. In the case of Aquinas, the teacher can be taken as an authoritative figure. She is not authoritative in the sense that the student learner should blindly accept whatever the teacher imparts, accepting it as authority even though the learner does not understand the precise steps by which the teacher arrives at a conclusion. But the teacher is an authority insofar as it is she who *guides* the student's reasoning processes so that the latter comes to cognize a given truth. For Aquinas, the student understands when she goes through the *same* reasoning processes as the teacher. As noted earlier, Aquinas states "Whoever teaches another *leads* him to the truth and so causes truth in his soul." The teacher is thus seen as someone who leads or guides the student, directing the student in the right paths so that she attains the truth.

In Locke, we find a rather different attitude towards the teacher and towards authority in general. *Locke's Conduct of Human Understanding* emphasizes a different kind of independence in the learner than that found in Aquinas. Locke, like Aquinas, emphasizes the need to use one's own reason to arrive at a conclusion. But he also has a less confident and more questioning attitude towards the human teacher. The human learner, at least after she has attained the use of reason, does not simply internalize the teacher's reasoning processes. Instead, she is someone who is cautious on all fronts, and questions on all fronts, whether it is her teacher, her own prejudices, or other forms of authority. She is more independent than the Thomistic learner insofar as she (not the teacher) should be her *own careful guide* in seeking knowledge. Her attitude is one of initial suspicion, if indeed not skepticism, towards any piece of information or claim imparted to her, regardless of the external source.[13]

This change in attitude towards authority, and by extension, to the teacher as an authority, can, I suggest, in turn be traced in part to a difference in the Thomistic and Lockean theory of knowledge. Recall that for both Aquinas and Locke, one receives, via the senses, sensible species or ideas from which one abstracts intelligible species or abstract ideas. The latter in turn can be used to formulate universal truths that then constitute the bedrock of *scientia* and knowledge. One difference between Aquinas and Locke concerns the epistemic significance of the intelligible species or abstract ideas. For Aquinas, when Mary's intellectual soul acquires the intelligible species of a dog, Mary *cognizes the real nature of the dog*. Aquinas is a metaphysical realist who holds that we have access to such real natures on the basis of the senses.

In contrast, Locke lived at a time when the new science was in the ascendant. Aquinas had largely accepted Aristotelian science, which saw the universe as constituted of various kinds of natural objects, each with its own essence or nature. Gassendi, Boyle, Locke and "the incomparable Mr Newton", were adherents of the new science, which saw the universe as above all mechanical—that is, as constituted of matter whose behaviour was governed by laws expressible in mathematical terms. For them, natural objects were constituted by corpuscles invisible to the human eye, whose configuration gave rise to the various perceptible features of the various natural kinds to be found in the physical world. Because of this, Locke held that we could never have access to the real natures or real essences of these various natural kinds. He writes in the *Essay*:

> Though the familiar use of Things about us, take off our wonder; yet it cures not our Ignorance. When we come to examine the Stones, we tread on; or the Iron, we daily handle, we presently find, we know not their Make, and can give no reason, of the different Qualities we find in them. 'Tis evident the internal Constitution, whereon their properties depend, is unknown to us. (Locke 1975, 444)

Thus, where the intelligible species provided the learner with, or at least was sometimes able to provide the learner with, epistemic access to the real natures of things for Aquinas, Locke holds that we can have no such epistemic access to ultimate real natures, at least with respect to the kinds found in the natural world. The real natures depend on the specific internal configurations of corpuscles imperceptible to the human eye, and as such must forever remain unknown.

In sum, then, where Aquinas held that one has in-principle epistemic access to, and could attain understanding of, every aspect of the natural world, there are fundamental aspects of the Lockean natural universe that are forever inaccessible to the human understanding. For Locke, then, the scope of human knowledge, of knowledge that humans could in principle attain, is reduced.

Moreover, this was likely not the only kind of reduction in the scope of human knowledge. Locke of course lived at a time of great intellectual upheaval. It was not only that the age-old and widely accepted theories of Aristotle about the natural universe were being overturned. The fact that theories that had been widely accepted for centuries could be overturned induced a further question: whether the more recent theories that were receiving increased attention—such as the atomism of Gassendi and Boyle, or Descartes' view which saw the material universe as a plenum—could be seen as providing a *true* description of the universe.

Locke's life-time witnessed the rise and increasing acceptance of the hypothetico-deductive method, which saw scientific theories as essentially *hypotheses* about the nature of the universe which had to be tested and confirmed. Commentators such as Alexander have argued quite convincingly that for Locke, the corpuscularian theory was precisely such an hypothesis, although one which might have good evidential support (Alexander 1997, 70). So the claim that the physical universe was ultimately corpuscularian might arguably not even have counted as *knowledge* for Locke. That it would not have done so becomes more likely in view of the skeptical questions raised by Locke's philosophical predecessors such as Descartes, Montaigne, and Charron.

Locke's *Essay Concerning Human Understanding* had as its aim the delineation of the 'certainty, extent and original' of human knowledge. What Locke accomplished there (and in other works) was to show that both the extent and the certainty of vaunted human knowledge were more circumscribed than previously thought. It is this significant reduction in the scope and certainty of what one could know in Locke's time that likely motivated Locke's suspicion towards all forms of received authority, including the Scholastically-based teacher. Such teachers had purported to guide one to knowledge about the ultimate natures of the natural world. But in view of the intellectual upheavals that were taking place, and the ensuing sense of uncertainty this engendered, such guidance did not seem wholly reliable. In his *Some Thoughts Concerning Education*, Locke thus recommended that children be tutored at home, so as to avoid schools which still propounded Aristotelian doctrines as providing the truth about the natural world.

In sum, the key difference between Aquinas' and Locke's theory of learning concerns the role of the teacher or instructor in that learning. Unlike Aquinas, Locke does not see the teacher as an authority who has the role of guiding the student's reasoning processes so that she attains *scientia*. Rather, the learner is to be suspicious of all forms of received authority, subjecting them to the scrutiny of reason. Locke's very different conception of the role of the teacher (or indeed any other authority) in learning stems, at least in part, from the intellectual upheavals of his times which suggested that what the human could know was much less, and perhaps much less stable, than had been previously thought.

4 Conclusion

In this paper, I have tried to show how acceptance of a particular theory of knowledge can impact one's theory of learning and one's view of the teacher's role in that learning. Locke and Aquinas share many points of similarity in their theory of knowledge which lead to similarities in their theory of learning. However, a crucial difference lies in their views about the scope and certainty of human knowledge. The reduced scope and certainty of human knowledge in Locke's time, and the shaken confidence in the (Aristotelian) teacher as guide led Locke to re-conceive the role of the teacher in the acquisition of knowledge.

In this connection, it is interesting to note that higher-education institutions are now increasingly emphasizing skills that enable life-long learning. The conditions that we now live in are not unlike those of Locke. With globalization, increased connectivity, and the surge in information come many competing theories and world-views, each vying for attention. What may have previously been accepted as received wisdom rapidly becomes obsolete. In this context, it is not surprising that the role of the university teacher is now arguably conceived more along Lockean lines. Such a teacher is not primarily seen as guiding the student by enabling the internalization of the relevant processes that enable the latter to accomplish learning. Rather her aim is to equip the student with life-long skills in independent reasoning, so that the latter can be her own careful guide in evaluating the various knowledge-claims that are thrust at her.[14]

End Notes

1. As some commentators have noted, this translation is not perfect. (See, e.g., MacDonald 1993, 162ff). This being so, I shall leave '*scientia*' as an untranslated term in this paper.
2. For Aquinas, the ultimate form of knowing is *intellectus*. However, the appropriate comparison in this paper is that between Aquinas' and Locke's accounts of *ratiocinative* knowing (i.e. knowing that involves reason). Hence the paper

will concern itself with comparing Aquinas' account of *scientia* and Locke's account of knowledge. My claims that *scientia* and knowledge are respectively the highest forms of knowing for Locke and Aquinas are thus qualified claims—viz., that they are the highest forms of knowing *through ratiocination*. (I thank the editors for help on this point.)

3. See Aristotle, *De Anima*, 3.7.431b2 in Aristotle (1968), 64; and Thomas Aquinas, *Summa Theologica*, I.12.2 (hereafter *ST*) in Aquinas (1981), 123.
4. This discussion involves a number of concepts (e.g. the sensitive soul, the intellectual soul, sensible and intelligible species) which may be unfamiliar to the contemporary reader. A thorough disambiguation and elucidation of these concepts would take this paper too far afield. Clear and well-rendered accounts of the process by which a thinker comes to grasp a real nature (and brief concomitant accounts of these specific concepts) are to be found in Macdonald (1993), 160–61, and Carriero (1990), 9–22.
5. This paper will not be concerned with the human's grasp of other real natures, such as those possessed by angels.
6. Philosophers like John Haldane have argued that Aquinas' account may be more complex, according a key role to the agent or active intellect in the formation of intelligible species. See Haldane (1992). In this paper, I defer consideration on the issue of the precise means by which the intellect acquires the intelligible species, given the presence of the sensible species in the knower. Instead, my broad claim here is that Aquinas sees sensory contact with the relevant objects as a *sine qua non*—and in that sense a foundation or starting point—for grasping their real natures. As I shall show later, he is, at least in this respect, not dissimilar to Locke. I thank Professor Haldane for his very helpful comments and discussion on this issue.
7. See, e.g., Aquinas (1956), Prologue.
8. See for instance MacDonald (1993), 185–88.
9. See Descartes' *First Meditation*, in Descartes (1996), 12–15.
10. The charge that Aquinas, as an epistemological optimist, had failed to consider skeptical arguments in respect of the pitfalls that may prevent the knowledge-seeker from attaining genuine knowledge, is quite common. Recently, writers like Stump (1991)and Kretzmann (1991) have argued that this charge is ungrounded. They argue that Aquinas was in fact an externalist and a reliabilist about knowledge. Once again, I will defer consideration on this reading of Aquinas, and will only note here that, even if one accepts this reading, an alternative version of the subsequent contrast that I draw between Aquinas' and Locke's theory of knowledge (and hence of learning) will still obtain. The claim now would be that Aquinas tends to assume the mechanism by which we gain knowledge is largely reliable, while Locke, given his adherence to the new science, has no such faith in this mechanism.
11. See Macdonald (1993), 174ff.
12. Aquinas, *On the Apostle's Creed*, Prologue, para. 5.
13. Aquinas does note that disputations have two purposes—one for teaching and one for removing error. In the latter form of disputation, too, Aquinas' views would be at odds with Locke's. Aquinas writes of the disputations for removing error:

In disputations of this sort you should above all use authorities acceptable to those with whom you are disputing; with Jews, for example, you should appeal to the authority of the Old Testament; with Manichees, who reject the Old Testament, you should use only the New; with Christians who have split from us, e.g. the Greek, who accept both Testaments but reject the teaching of our saints, you should rely on the authority of the Old and New Testaments and of those church teachers they do accept. And if you are disputing with people who accept no authority, you must resort to natural reasons. (*Quodlibet* IV. Cited in Mooney and Nowacki 2011, 109–110.)

The procedure in this latter form of disputation is different from that of Locke's. Locke emphasizes using one's reason as a careful guide in *all* circumstances, and does not think the learner should accept *any* assumptions as given. Aquinas position here differs from that of Locke's in two ways. Firstly, Aquinas recommends in the first instance that the disputer should assume the other person's own position, and try to convince the latter *using the latter's given assumptions*. It is only as a last resort that she should appeal to natural reasons. Thus, Aquinas thinks that the process of the other person's coming to accept the particular position need not involve the questioning of the assumptions she takes as given. Secondly, the passage indicates that in embarking on this latter form of disputation, the main aim is to guide this person (by the various means suggested) towards the truth. This once again indicates that for Aquinas, it is the disputer who is in possession of the truth, and that her aim is to guide the person in question to that truth.

14. This paper was read at a conference on Aquinas on Education and the East, held at the Singapore Management University, 15–16 October 2009. I thank the participants and audience for their very helpful comments on the paper. I also thank the editors for their careful and detailed comments which have resulted in a much improved paper.

References

Aaron, R.I. 1955. *John Locke*. Oxford: Clarendon.

Alexander, P. 1997. Boyle and Locke on primary and secondary qualities. In *Locke on human understanding: Selected essays*, ed. I.C. Tipton. Oxford: Oxford University Press.

Aquinas. On the Apostle's creed. Available at: http://dhspriory.org/thomas/Creed.htm. Accessed 2010.

Aquinas. 1953. The teacher (Q.11). In *Truth*. Trans. James V. McGlynn. Chicago: Henry Regnery (cited as *DM*).

Aquinas. 1956. *Exposition of the posterior analytics of Aristotle*. Trans. Pierre Conway, O.P. Quebec: M. Doyon.

Aquinas, Thomas. 1981. *Summa Theologica, complete English edition in five volumes*. Trans. Fathers of the English Dominican province. Allen: Christian Classics.

Aristotle. 1968. *De Anima, Books II, III*. Trans. D.W. Hamlyn. Oxford: Clarendon Press.

Carriero, J. 1990. *Descartes and the autonomy of human understanding*. New York/London: Garland Press.

Cranston, M. 1957. *John Locke: A biography*. London: Longmans.

Descartes, R. 1996. *Meditations on first philosophy*. Trans. and ed. John Cottingham. Cambridge: Cambridge University Press.
Gilson, E. 1948. *History of philosophy and philosophical education (Aquinas lecture 1947, fall)*. Milwaukee: Marquette University Press.
Haldane, J. 1992. Aquinas and the active intellect. *Philosophy* 67: 199–210.
Kretzmann, N. 1991. Infallibility and ignorance. In Bosley, R. and M. Tweedale eds. 1991. Aristotle and his medieval interpreters. Canadian Journal of Philosophy, Supplementary Vol. 17 (Calgary: University of Calgary Press).
Locke, J. Some thoughts concerning education. Available online at: http://www.fordham.edu/halsall/mod/1692locke-education.html. Accessed 2010.
Locke, J. 1975. In *Essay concerning human understanding*, ed. Peter H. Nidditch. Oxford: Clarendon.
MacDonald, S. 1993. Theory of knowledge. In *The Cambridge companion to Aquinas*, ed. N. Kretzmann and E. Stump, 160–195. Cambridge: Cambridge University Press.
Mooney, T.Brian, and Mark Nowacki. 2011. *Understanding teaching and learning*, St Andrews studies in philosophy and public affairs. Exeter: Imprint Academic.
Stump, E. 1991. Aquinas on the foundations of knowledge. In Bosley, R. and M. Tweedale eds. 1991. Aristotle and his medieval interpreters. Canadian Journal of Philosophy, Supplementary Vol. 17 (Calgary: University of Calgary Press).
Tarcov, N. 2003. Enlightenment liberalism (2nd section). In *A companion to the philosophy of education*, ed. R. Curren. Oxford: Blackwell.
Wolterstorff, N. 1996. *John Locke and the ethics of belief*. Cambridge: Cambridge University Press.

Part III
Education and the East: Reflections on Educational Policy

Reorganising Schools as Social Enterprises: Play Schools and Gifted Education

Soo Meng Jude Chua

1 Introduction

In a recent pamphlet titled *Mutual Futures: Ed Balls, Michael Gove and the Challenge to Faith Schools,* Francis Davis and Nathan Koblintz suggest that Catholic faith schools in England and Wales be reorganized as social enterprises (Davis and Koblintz 2009, 6).[1] Sustained by a centralized mutual fund, the "Catholic education sector …would be driven by inclusion, [and] social innovation […]. Complimentary initiatives such as credit unions of 'banks for the unbankable', language training for migrants, businesses and social enterprise advice could also be housed within the resources of the mutual… [and] school campuses would be designated… as 'social enterprise' zones or social silicon valleys forming local hubs from which new institutions could be launched or renewed" (Davis and Koblintz 2009, 11).[2]

This policy proposal was aimed at helping Catholic schools survive massive hidden taxes by the British Government. This taxation initiative had threatened to bring about the closing of several Catholic schools. By reorganizing themselves as social enterprises, Catholic schools would have access, via the common mutual fund, to profit-making dimensions of the social enterprise that could help defray expenses. In addition to the attractive economic ramifications of the policy proposal, reconfiguring schools into social enterprises may be expected to have other welcome effects. In what follows, I will focus upon non-economic educational and social reasons, and consider the support they could offer for this policy. In sum, I will argue that social enterprises need not be merely useful *economic* policy

An earlier version of this paper was read at the "Aquinas, Education and the East" Conference at Singapore Management University in October, 2009. My thanks especially for helpful comments by John Haldane, John Ozolins and other participants. I am also grateful to John Finnis for comments on an earlier draft.

S.M.J. Chua (✉)
Associate Professor at National Institute of Education,
Nanyang Technological University, Singapore, Singapore
e-mail: jude.chua@nie.edu.sg,

T.B. Mooney and M. Nowacki (eds.), *Aquinas, Education and the East*,
Sophia Studies in Cross-cultural Philosophy of Traditions and Cultures 4,
DOI 10.1007/978-94-007-5261-0_10,

technologies promoting financial stability and freedom, but rather by virtue of their very nature as *educational* policy technologies can lead to desirable educational outcomes for students, positive professional development opportunities for teachers, and positive social spillover effects for the wider community.

This paper engages two groups of stakeholders. First, those who, as members of the Catholic faith community, are concerned with both education and the broader range of social concerns within which they find themselves. Second, those who, while not part of the Catholic faith community, are nonetheless stakeholders in the community within which the Catholic social enterprise zone is situated. Sharing the Catholic justificatory framework can, I hope, go some way to building mutual trust and support between these two stakeholder groups. For simplicity, I will focus on those justifications internal to the Catholic tradition and assume that, *mutatis mutandis*, the desirability of the social initiatives can be made apparent to the wider non-Catholic community.

In the following sections I consider ways of conceptualizing school Social Enterprise Zones (hereafter SEZ) so that they can be seen as places where spiritual education and moral education can take place and find support. In Sect. 2, I draw on Aquinas' scholastic metaphysics and suggest that participation in SEZ-level projects provides opportunities for teachers and students alike to share in God's 'creative play.' This participation in the Divine playfulness results in what we may properly call 'play schools.' Students at play schools assimilate the playful idea that social action is not only an overflowing of the spiritual but is itself a worthy object of spiritual contemplation. In Sect. 3, I consider how being actively part of an SEZ fundamentally changes the epistemological situation of students and has the potential to open them up to new moral insights. I call this awakening of the moral sense of students 'gifted education'. Such awakening inevitably leads to a fruitful challenging of one's normative biases and prejudices. Finally, after exploring the spiritual and ethical educational dimensions of SEZ, in Sect. 4 I offer a theologically-motivated suggestion for further research within the Catholic tradition that would explore how gifted education in play schools can help us cultivate a capacity to see as Christ sees so that we become better fit to respond to our neighbor as a 'concretized Christ' in each time and place, thus revealing Him in history.

2 Play Schools

I now explore the notion of divine play which will be taken up in the notion of a play school. To understand divine play, it is helpful to begin with Aquinas' account of God as pure Existence, or more accurately, Be-ing, *Esse*.[3] Recent work by scholars tracing the neo-platonic sources of Aquinas' metaphysics acknowledge that Aquinas retained the Pseudo-Dionysian insight that God is really a kind of intensive energy (*virtu*), dynamic and active.[4] Aquinas employs the Aristotelian notion of 'actuality' to express this idea: God is Pure Act or Perfection, the absolute source of existence (*esse*). More than that, God's pure actuality is unlimited by any determining principle (Aquinas 1968, 54–62). By way of contrast, for any creature, *what it is* is really distinct from *that it is*, meaning that its essence is really distinct from its existence.[5] A creature receives its existence (*habens esse*) from God in the degree determined by the receptive capacity of its essence.[6]

God's activity, like his very existence, is not constrained by any extrinsic limiting principle. Any act of God, even his creation of creatures, is a purely free act.[7] Further, God is perfect within himself and needs nothing else to be perfect, and so when he creates, there is no benefit accruing to God in the act of creation. Any benefit accruing from creation must therefore fall on the side of the creature rather than on the side of God. But that which is done freely and purely for the benefit of another is an act of love and therefore a *gift* (*caritas*). This gifting can also be described as *play*, as Hugo Rahner suggests:

> When, wherefore, we speak of God the Creator 'playing', there lies concealed in that phrase the metaphysical truth that the creation of the world and of man, though a divinely meaningful act, was by no means a necessary one so far as God himself was concerned...[The playing of God, God's creative activity] is too full of a most profound meaning to be anything but immensely serious—but we do, in so conceiving it, avoid falsifying its quality or presenting it as an act which flowed from God's nature as a metaphysical necessity, as though God were in some way subordinate to his own works or, in some pantheistic sense, identical with them. God is free. Out of the vast multiplicity of possible ones he calls one particular order into being. (Rahner 1963, 11–12)[8]

Now, SEZ are places where one would find many such gifting-playful activities. I will first list these various strands of play before I suggest their educational relevance.

Firstly, we may assume that all SEZ activities will have an implicit commitment to moral standards, a commitment which may itself be expressed as one layer of play. To the extent that an SEZ's operations and activities are guided by the principles of practical reasonableness, they partake in what may be called *moral* play. This notion of moral play is latent within John Finnis' *Natural Law and Natural Rights*:

> The requirements of practical reasonableness (which generate our obligations) have a 'point' beyond themselves. That point is the *game* of co-operating with God. Being *play*, this co-operation has no point beyond itself [...] [I]f we ask why God creates, no answer is given except the one implicitly given by Plato: play—a free but patterned expression of life and activity, meaningful but with no further point [...] [E]ven one who goes beyond Plato to accept that man is called to a friendship of devotion to God will grant that such friendship takes the form of sharing, in a limited way, in the divine play. Practical reasonableness, therefore, need not be regarded as ultimately a form of self-perfection. This is not its final significance. Nor, on the other hand, are its requirements sheer categorical imperatives; they gain practical force from the basic explanation that can be provided for them—that they are what is needed to participate in the game of God. (Finnis 1980, 408–409)

For Finnis, the 'serious life' or the life guided by one's moral obligations is a kind of play *with God*. This is because if creation and the principles of practical reason are God's playful work, then our moral life is simply an unfolding of that. It is as if God invented a game, freely devised its rules and then called us into it. The man who lives the serious life is God's playmate. Aquinas' statement that the natural law (i.e., the first principles of practical reason) is a certain participation of God's own will called the Eternal Law captures this.[9] The morally worthy life, the serious life, is ultimately a participatory extension of God's own playfulness. With this Platonic perspective Finnis introduces play as the new and further point of one's serious life: one takes one's moral obligations serious for the sake of playing (with God).

Now, there are two ways to think about this. Firstly, we can think about moral absolutes and their implications prescribing what *not to do,* which we obey. These prescriptions of practical reason are the result of a *rigid,* deductive unfolding of negative rules—albeit rules which issued from a God who played when he created the world and determined its providential order. Here in one sense the practically reasonable man participates in God's play: by the faithful *continuation* of a choice that *had once* freely emerged. The metaphor here is the *extrapolation of a straight line* that had been freely extended from a point. However, this extrapolated line's claim to 'play' is actually only through its association with an earlier straight line that had itself been playfully drawn.

But there is a second way we can think about our moral participation in God's play. Think about the many free choices[10] we will make when we decide what *to do,* at the level of individual agency or at the level of institutional decision making, such as choices before incommensurable options, or in acts of *determinatio,* etc.[11] Here the metaphor is the ongoing drawing of *fluid curves* and other freely designed patterns on paper: each new curve shares in God's play through having in itself that playful quality. Here, man at play (*homo ludens*) participates in God's play through his own creative play, and not just through the dogmatic and faithful perpetuation of something *once played.*

If we look more closely we can detail these modes of moral play which are determinate ways we do good in SEZ. For a start, unlike explicitly profit-only driven enterprises which are focused on maximizing one type of good, social enterprises can be open to a plurality of dimensions of worth, or a plurality of goods, and these goods are quite often *inter-alia* incommensurable. The result of this is that choices to address and choices to remain committed to addressing[12] *this* set of goods or dimensions of worth rather than *those* are freely chosen. No computational procedure can determine *the* way forward. The choices which determine which goods the organization pursues are settled by free choices. In this one sense, such choices are playful-gifts—freely given, freely made, freely determined by and only by the organization's agency. Play in this organizational sense, which we might label *social-entrepreneurial* play, therefore permeates the SEZ.

Again, there can also be another more specific kind of play, *managerial* play, which can also be found within the SEZ. Particularly for Catholic schools that are seeking to escape the "terrors of performativity"[13] and are ongoingly catechized[14] with what organizational theory calls 'policy technologies of foolishness,' i.e., activities that can help participants escape normative biases and prejudices, it is not difficult to imagine that SEZ become very dynamic places where participants (employees, students, leaders) are encouraged to think playfully afresh. For example, during workshops or retreats participants may be given opportunities to explore or revisit classic religious texts and ideas and the humane ideals they represent, and be welcomed to draw upon these ideals when they construct their own reflective speeches (*oratio*). Some of these ideas can emerge at professional sharing sessions while others might be published in school newsletters or professional journals, etc. In this way, participants experiment with new Christian identities. Experimentation is tentative, non-committal, fluid, free, open to change…in a word: playful. Even when participants settle on a particular Christian identity, one need not imagine that such play will end: organizational life is an ongoing joust with performative pressures, and there is always room for ongoing repentance and re-conversion.

There is also *curricular* play. This is the sort of play that teachers engage in their professional capacities as curriculum designers. Such play is seldom if at all linear, but rather experimental and exploratory, and to a great extent unpredictable and open to embracing that which is surprising. At the classroom level, lesson planning and curriculum design activities can and should be playfully—or "goal-less-ly", to borrow an expression from Herbert Simon—open to assimilating as new goals and strategies those welcome consequences that emerge often unintentionally during the design processes (Simon 1997). For instance, I knew a teacher whose lesson on the science of climate change was interrupted by students who with great excitement began to raise ethical concerns about global warming and that led her to re-design her lesson to incorporate an ethical discussion into her unit plan in order to make the lessons more engaging.[15]

Finally, there must certainly be *scholastic* play by students. Like Aquinas, we could think of scholarly learning as a kind of play. As he expresses it in his commentary on the *De Hebdomadibus* (*Expositio libri Boetii De ebdomadibus*) contemplation may be suitably compared to play for two reasons. First, contemplation is like play because it is delightful; and second, contemplation is sought for its own sake and need not be ordered to anything else, just like play (Aquinas 2001, 5). Building on that idea, one can also foresee students in SEZ involved in social projects learning to make sense of social issues, and trying to interpret complex real life problems. While not 'contemplative' in nature, such practical thinking, involving sense-making, is often abductive. It is hence, as Susan Petrilli explains, "risky…[as] it advances through arguments that are tentative and hypothetical…[it is] regulated by the law of creative love…" (Petrilli 2007, 117–130) and for that reason also playful.

I have just shown that when schools are organized into SEZ, there is clearly a proliferation of play. I listed these as: *moral* play, *social-entrepreneurial* play, *managerial* play, *curricular* play and *scholastic* play. I now suggest how this reality can be educationally beneficial. As I pointed out, present in these 'play schools' is *moral* play, which makes sense of all our moral endeavors in SEZ as a kind of play with God. Furthermore, the various other strands of play, viz. *social-entrepreneurial* play, *managerial* play, *curricular* play and *scholastic* play, are all determinate, modal expressions of positive moral play, of our *doing good*. This implies that one's play in whichever of these strands in SEZ can also be conceived as a sharing of God's play.

Now, such an idea can be communicated to school leaders, administrators or teachers through professional development courses and to students within a moral or spiritual education curriculum, and be made an object of their reflective contemplation. For those belonging to the Catholic tradition particularly, this may be very desirable. Such *reflection on* 'playing with God', or on God generally, may also be characterized as the realization of the image of God (*imago Dei*) in the student and teacher. This is because we do end up thinking about God. Thus the reflective person conforms to God, who also thinks through and about Himself. This way of imaging God is what the Thomistic scholastic tradition considers the most excellent form, 'the image of conformity.'[16]

3 Gifted Education

Some theologians and philosophers who dislike 'metaphysics' may reject my arguments above. My description of God as playful, unlimited Being would for them be wrong. They would prefer theories about God to be 'phenomenological.' Jean-Luc Marion explains what this involves:

> [I]n phenomenology—that is to say, at least what it intends, in the attempt to think in a non-metaphysical mode—it is a question of showing. To show implies letting appearances appear in such a way that they accomplish their own apparition, so as to be received exactly as they give themselves. (Marion 2002, 7)

Such a phenomenology is contrasted with the metaphysicians' construction of conceptual 'idols.' For Marion, conceptual ideas become idols when we project our own cognitive constructions at the object of our investigation, and therefore gaze at our own inventions. Thus any metaphysical description of God, which is built from concepts, actually sets God aside and substitutes something else of our own craft and making in His place. In *God Without Being,* Marion chides as suspiciously idolatrous Aquinas' metaphysical naming of God as Being (*ens*):

> One must choose: if theology proceeds by the apprehension of concepts, as a 'science,' then, for it [...] the *ens* will be first, and man's point of view normative [...]. But if theology wills itself to be *theo*logical, it will submit all concepts, without excepting *ens,* to a 'destruction' by the doctrine of divine names, at the risk of having to renounce any status as a conceptual science [...] [Such a choice] Saint Thomas did not make, the Saint Thomas who pretended to maintain at once a doctrine of divine names and the primacy of *ens* as first conception of the human understanding [...] [Yet] the claim that the *ens,* although defined starting from a human conception, should be valid as the first name of God...does not easily escape the suspicions of idolatry. (Marion 1991, 81–82)[17]

Marion has in his own way retracted his criticisms of Aquinas.[18] Even with Marion's retraction it is still difficult to take sides in this matter. It has been suggested that Marion's *retractio* excessively and wrongly downplays Aquinas' belief in the reliability of metaphysical concepts for theological thinking.[19] Meaning to say, Marion's criticisms may still hold true for Aquinas if Aquinas is properly understood. One might even argue that when leaving behind Aquinas' metaphysics, one achieves the true spirit of St Thomas. This is something John Caputo's interpretation of Aquinas' own theological experience could suggest. Caputo reads in the early Aquinas an articulate metaphysician and also a later 'silent' Aquinas who declared his metaphysical work to be like straw, and so put it aside. Caputo suggested that Aquinas had a mystical experience of God and after that downplayed his metaphysical works in preference for that moment of God's self-disclosure.[20]

In any case, I am eager to avoid alienating readers who share Marion's reservations regarding onto-theological thinking and who are inclined to phenomenology. While I do not recant my arguments in Sect. 2, here in Sect. 3, I explore ways SEZ are educationally beneficial without drawing on Aquinas' metaphysics. I suggest how possibly *Mutual Future's* policy recommendations result in what I call 'gifted education', which is the awakening of the moral sense of students.

To do this, I draw on Marion's phenomenological work. Marion's works call to our attention "saturated phenomena." These saturated phenomena refer to phenomena which overflow our epistemic capacities. Here, there is more light than the eye can behold; our vision is overwhelmed. There is too much which the phenomena give to the subject, and whilst the subject is given in such an excess, he is bedazzled, and since he cannot take in all that has been given, his initial experience is blindness and a lack of understanding: "the given intuition overwhelms our capacity for reception [...] [A]n excess of unforeseeable intuition floods our intentional horizons, fills them, saturates them, and overflows their limits. More is given than can be received" (Miller 2008, 86–87). Phenomenological openness to saturated phenomena in Marion is "a style of thinking, a rigorous practice, not the workings of a system. It entails a process of perceptual and intellectual purgation in which the phenomenon, any phenomenon, is considered in its sheer givenness or self-presentation" (Robinette 2007, 89).[21] The task in phenomenology is to find ways to appreciate saturating phenomena, which entails not dismissing phenomena simply because one initially cannot grasp them.

Robinette has recently considered the suffering around us as a kind of saturated phenomenon (Robinette 2007). Tagging on Robinette's insights, I think SEZ can lead through 'faith' to a phenomenological openness to the saturated phenomena of suffering.[22] Robinette suggests that saturated phenomena need not be narrowly confined to positive phenomena; rather, there are experiences of saturated *negative* phenomena: "the experience of *radical negativity* might well be the kind of saturated phenomenon that when more thoroughly analyzed displays a dynamic much more dialectical in character" (Robinette 2007, 101). Saturated (negative) phenomena include what is encountered in "negative contrast experiences" (Schillebeeckx 1980a, 816–819), as Edward Schillebeeckx calls them. These are experienced in confrontation with grave meaningless suffering, and from which protestations arise. "Whereas the aesthetic encounter is enrapturing, 'goal-less', and playfully expansive; suffering touches off a critical, cognitive force for its overcoming" (Robinette 2007, 102). In the negative contrast experience, the person grasps that what he is seeing or experiencing simply 'should not be!' But this is not mere protestation based on a judgment grounded in principles one already grasps with full presence of mind; instead these principles may be psychologically latent, but now in that experience are fore-grounded. In negative experiences of contrast the agent is transformed epistemologically and praxeologically: his normative frames are challenged, and he becomes psychologically conscious of practical principles which now point out new judgments and actions. Schillebeeckx goes so far as to suggest that new ethical insights which challenge conventional value-frames arise after negative experiences of contrast, and he downplays the role that speculative thought has for generating new ethical paradigm shifts (Schillebeeckx 1969, 153–154).

Like positive saturated phenomena, the intentional direction in negative experiences of contrast seems reversed. Instead of the agent constituting the data, the experienced data *constitutes the agent.* This it does through contradicting, challenging and eventually revising the agent's ethical frame: the epistemic terms and conditions with which he initially (e)valuates the data (as something acceptable or even positively welcomed) is rejected; now for him, 'this cannot be!' Again, in negative contrast experiences, much like encounters with positive saturated phenomena, the agent experiences the "blinding bedazzlement"—the ethical demands which the data (viz. the grave suffering) impose upon him are initially incomprehensible, in the sense that they are not consistent with or exceed his current frame. "Must we not also speak of *negative* bedazzlement, the fragmentation of horizons, the traumatizing of a language that cannot properly name the radical mystery of suffering and evil?" (Schillebeeckx 1969, 153–154). Indeed, bedazzled, and blind, he may *try to see*; he may begin to search for premises and arguments *to corroborate or warrant these new ethical demands.* Thus Schillebeeckx offers the sociological observation that these ethical insights which derive from experiences of contrast often begin the speculative development of supporting theory or general principles (Schillebeeckx 1969, 153–154). Furthermore, the experience of contrast reveals itself on its own terms, and not based on the subject's epistemic terms: if the subject clings on to *his* frame (with its premises, values, epistemic criteria etc.), the experience with its ethical intuition is resolved and reduced into that which is unanalyzable, un-warranted, un-quantifiable, un-defensible. For the data to fully disclose itself one must obey *its* (epistemic) *rules,* the conditions *which it sets.* The metaphor here is the gift, which one receives without one's determination. In the economy of exchange, one is willing to take the offer when the condition *one sets* has been satisfied. Thus if you and I are to exchange books, I would accept your book only on the condition that you give me *that* one, a condition that I determine. In giving on the other hand we do not find such determining conditions: when you give me something, it is not for me to determine the reasons for which you give that to me. Rather it is fully your initiative, and your conditions determine the gift. It is given *as you give it.* Put in another way saturated phenomena set the conditions for their own disclosure; they *give* and therefore *show* on their own terms. Marion describes their *anamorphosis* using the analogy of a person gazing at a painting:

> Though here it is not a matter of pictorial procedures, these can help us clarify by analogy the phenomenological sense that I am imposing on this term [*anamorphosis*]. This procedure involves first presenting to the uncurious gaze of the viewer a surface entirely covered with colored pigments but apparently void of any unrecognizable form whatsoever, then moving this gaze to a precise (and unique) point from which it will see the

> de-formed surface trans-form itself in one fell swoop into a magnificent new form [...]. To accede to it, not only must a gaze know how to become curious, available, enacted, but above all it must know how to submit to the demands of the figure to be seen: find the unique point of view from which the second level [new] form will appear, therefore make numerous and frequently fruitless attempts, above all admit that it would be necessary to alter one's position (either in space or in thought), change one's point of view—in short, renounce organizing visibility on the basis of the free choice or the proper site of the disengaged spectator, in favor of letting visibility be dictated by the phenomenon itself, in itself. (Marion 2002, 124)

As one form of saturated phenomena, albeit *negative* saturated phenomena, negative experiences of contrast are revealed and received on their own terms, and not on the subject's terms. "The phenomenon no longer appears as soon as I open my eyes to it, like an object summoned to the gaze that produces it; rather it arises when my gaze has satisfied the demands of the perspective, therefore of the appearing, proper to what shows itself starting from itself" (Marion 2002, 124). Indeed, the subject is challenged to revise his or her frames, to modify his or her terms of engagement. In this way, experiences of negative contrast are experiences of *given-ness* [*donation*]. Negative experiences of contrast are thus *given,* and subjects who experience such contrast experiences are *gifted.*[23]

Now, SEZ can clearly be the consequence of such gifted-ness. Indeed social enterprises quite often are *social* enterprises precisely because they respond to existing unjust suffering and seek to address this. Social entrepreneurship has often resulted from negative experiences of contrast: 'this cannot be! We need to do something about it.'[24] Nevertheless, *the causal connection is equally plausible in reverse:* when reconstituted as SEZ with an ethos of open responsiveness to surrounding suffering, schools can be places *for education in gifted-ness*. Gifted education here refers not to the cultivation of the intellect for IQ tests but the nurture of the phenomenological disposition, which is that willingness to engage saturated phenomena. In this case, students are encouraged to encounter the negative saturated phenomenon of suffering which when received as 'gift', results in negative experiences of contrast. When schools are constituted as SEZ, the presence of an institutional commitment to social entrepreneurship can itself *lead to* such gifted-ness. Let me detail the reason.

A saturated phenomenon, we recall, is received in initial blindness. One's original frame is at odds with the given, and so cannot capture it well. It is not 'visible'; there is 'nothing there': there is no issue, no problem, no harm, no disaster, no injustice. In persistently engaging the given *qua saturated phenomena,* as 'something-there-which-has-more-to-give-than-I-can-receive', one does so only through acknowledging, in this blindness, that there is more than I can see. Now take note: this acknowledgement is not epistemologically warranted *a priori* at that point in time when one decides to persist in looking. Instead it is warranted only *a posteriori,* when the given shows itself and crosses in contradiction one's frames, and when these frames have been revised.

So, at the point when one cannot see something there, it appears possible that there is *really* nothing to grasp, as much as it is possible that there is something to grasp, but now escapes my sight because it saturates it: "Prior to my receiving the given so as to make it visible, how would I see that I do not see? Or how would I distinguish between blindness of intuitive excess and that of intuitive lack?" (Carlson 2007, 160). Hence when faced with the choice to dismiss the data in the light of one's current frames (e.g., "Perhaps there's nothing wrong here; it's a little warranted sacrifice in the interest of maximizing the majority's preferences…") or to attend to it, persist in reception and be willing to change one's frames (e.g. "This can't be right; perhaps the consequentialist thinking which warrants this needs to be challenged in the light of these intuitions…"), it appears that either way forward involves a gamble, an act of willed faith. Conversely, *if there is something of a saturating phenomenon there,* then unless I had such a faith, or such a will, I would not be able to see it. Thomas Carlson speaks therefore of a second form of blindness, which issues not from one's lack of capacity to see, but over and above that, from one's unwillingness to see:

> [This] blindness would result not so much from the *finitude* of my capacity to see the excess at which I nonetheless look, but indeed more from that in me which resembles an *infinitude:* namely, from my *will,* where I can prove not simply unable but indeed unwilling to see or receive the given, refusing to be constrained by any obligation or imperative at all. (Carlson 2007, 160)

The implication of this is that first and foremost what is epistemologically and phenomenologically useful before saturated phenomena, is the act of faith which turns the will from this second blindness, before it can proceed to cure the first blindness. It is, as Carlson sums up, a case of *fides quarens intellectum* (faith seeking understanding):

> 'Seeing is believing,' as the saying goes, which here means not, as one Thomas might think, that belief ensues from the force of evidence seen but rather, indeed, that I must first believe, thanks to a decision made in darkness, in order to see at all. This decision to see, then, might be understood to fall squarely within the logic of a faith seeking understanding, which means that the possibility of phenomenology itself relies on a necessary pre-phenomenological movement of faith. (Carlson 2007, 162)

This 'faith' can be supplied by the institutional will in SEZ to not dismiss those negative saturating phenomena of suffering. Social entrepreneurship's ethos, which is the orientation and stance of *social caring* through a directed *searching* attentiveness to negativities ("*ecce homo*!—but can this be right?"), and moral protestation ("perhaps this should not be? Perhaps there's something really unjust here?"), is precisely the infusion of this 'faith', of this persistent will to see. In this way SEZ encourages those originally indifferent to press onwards trustingly in blindness ("let's research and develop the theory as we go") and not by sight. Thanks to this 'faith', one is enrolled in SEZ for gifted education. Otherwise, one passes the suffering by: "Is there a wrong? It's not entirely clear—perhaps it is as it should be. *Perhaps.* Let it be then…"[25]

In other words, SEZ supports the education of a socially and morally aware generation, who would be willing not only to address suffering, but who are also willing to challenge paradigms and discourses that make nothing of these sufferings.

4 Memoria Christi: Final Thoughts

I have for the most part spoken of the educational benefits of SEZ for teachers and students. But I also glimpse in SEZ educational benefits for potential or current religious ministers. After all, one may imagine in some of these 'play schools' persons who have completed or are completing theological or pastoral studies with a view to ministering within the SEZ. In this last part of my essay, I wish briefly to explore how being located in SEZ can support their pastoral formation. This I do by painting in broad strokes what implications 'gifted education' could have for one's theological appreciation of God's self-revelation through Jesus Christ. I do scant justice to these ideas, and they welcome further research and discussion, to be sure.

We have been talking about the engagement of suffering that leads to new moral sensibilities in SEZ and describing that as 'gifted education.' Still, gifted education is not just gifted education, period, but at the same time useful preparation for a fruitful ministry of Christian remembrance, leading ultimately to the showing of Christ. When participants are gifted, i.e. phenomenologically open to negative saturated phenomena leading to ethically responsive *praxis,* only then, according to Schillebeeckx, is possible *Christian anamnesis* or a critical *memoria Christi* (remembrance of Christ) in the sense that one performatively recalls the past:

> Like any living remembrance of the human history of suffering, Christian *anamnesis* or remembrance of Jesus' particular course of suffering develops a particular critical epistemological force. In that case however, its rebellious or challenging character does not lie in a theoretical remembrance of a past event, nor even *directly* in the articulation through proclamation of the suffering of Jesus, but rather in what the Bible calls 'remembrance'. There, for instance, it is said of God that he remembers his earlier saving acts by now bringing new saving acts to pass in the present. Reference to what is actually done here and now is an essential part of the biblical view of memory [...] [T]he church is a critical *memoria Christi* to the degree that its particular way of life can be shown to and is visible to all, presenting a challenge and leading to revolution, and in this respect is a living remembrance of Jesus who overcomes suffering. (Schillebeeckx 1980b, 820)

Instead of rote learning and recitation from the Text/Book, here is *memoria* that is a *performative* reconstruction of the past. This is not achieved however by merely "imitating what Jesus did," (Schillebeeckx 1980b, 820) in the past but rather by authentic participation in his general mission to resist and overcome negativities as they are found in the forms today. Moved by particular ethical imperatives surfaced through specific experiences of negative contrast in the face of particular contemporary forms of suffering, participants who resist these sufferings recollect for us the Lord Jesus, who did the same in his time. Now each instance of a recollected Christ is one possible facet of the Lord Jesus Christ had he been incarnate in this point in history. Each and every gifted response to negative saturated phenomena is one aspect of the disclosure of the full 'face' of Christ across history. Through these various ongoing remembrances, Christ is continually being further disclosed in multiple ways, or indeed, possibly infinite ways given the infinite contexts, into the future. In this way, through those who have 'faith' and are gifted (synonymously: graced), the policy to organize schools into SEZ may yet culminate in an overflowingly abundant Christology that continues to disclose *via* Christian *anamnesis* the Lord Jesus Christ, until He comes again. Ministers and pastors who are formed within these SEZ will therefore be part of this ongoing remembrance and revelation of Christ in our current time, in the current context.

End Notes

1. Also see Davis (2009a, b).
2. My italics. Francis and Nathan also give a handful of examples of Christian social innovation in Appendix A of *Mutual Futures*.
3. The key text here is Aquinas (1968).
4. See especially O'Rourke (1992, 133–188). Also Clarke (1994, 65–88).
5. See Wippel (1984, 107–132).
6. See Chua (2000).
7. Also see Burrell (1990).
8. Also see Boland (2007, 142), which cites Aquinas' *ST* I.44.4 *ad* 1: "To act out of need indicates that an agent is imperfect, that its nature is to act but also to undergo. This is not fitting for God. And so God alone is completely free, *maxime liberalis,* because God acts not for his own gain but solely from his own goodness." Other traditions contain interesting parallels to the idea of divine play: consider, for example, the Islamic Sufi notion of *lila* and the Hindu and Buddhist notions of *maya*.
9. See Finnis (1980, 398–403).
10. See Finnis (1994, 146–151); and Finnis (1997, 215–233). Also see Boyle (2002, 123–141).
11. Compare *determinatio* with Herbert Simon's satisficing, which seeks one 'good enough' solution amongst a possible plurality of options wherein the theoretical best is not available. See Simon (1997).
12. See Alkire (2000, 102).
13. See March (1978, 587–608, 1991, 71–87, 1994, 237–240, 1999a, 308–324, 1999b, 225–228). See also Chua (2009, 159–167).
14. See March (1999b, 225–228).
15. For more on this theme, see Chua (2008, 18–23).
16. On *imago Dei* see Cessario (1996, 43–48), and Aquinas (1995, 32–34).
17. However, Marion later recants this, explaining on Aquinas' behalf that for Aquinas, God is *esse* in so wise as to be infinitely distant from the *ens* that creaturely beings are, and hence in Thomistic metaphysics, the conceptual *ens* nowhere nears a sign to represent or frame God, *Esse*. Aquinas names God not *ens* (which is better reserved for creatures), but rather *Esse*, which is one of the participated—and hence constituting—principles of *ens*.
18. Marion (1991, xxiii): "[Aquinas] does not chain God to Being because the divine *esse* immeasurably surpasses (and hardly maintains an *analogia* with)

the *ens commune* of creatures, which are characterized by the real distinction between *esse* and their essence..."

19. See Shanley (1996, 617–625).
20. See Caputo (1982, 246–287).
21. See also Rivera (2009, 1–10).
22. One must be on the guard, however, against whimsically mis-ordering the way pedagogical aims are prioritized in schools in the education of children, and it is not wrong to consider as a priority those educational goals which immediately support the intellectual, psychological and moral formation of children appropriate to their age, and feature educational programmes for social justice education only to the extent that these are appropriate or when they do not detract from what *as schools* these social enterprise zones need to supply, namely a proper education and formation aside from education in social justice issues. Hence it may be probable that social justice education and concerns in these enterprise school zones might be more appropriate to students at the higher levels. In Singapore, for instance, only when they reach secondary schools do student teenagers embark on "service learning". For an argument why this ordering of concerns is necessary, see Chua (2006, 56–62). Thanks to John Finnis for alerting me to this concern.
23. C.f. Marion (2002, 266); see also Horner (2005, 116–117); also Rivera (2009, 1–10).
24. For instance, Mohammed Yunus' development of micro-credit banking evolved from his repugnance for bankers who refused to lend such small amounts to the poor. He considered this a grave injustice.
25. For a related study of how photography can be employed to promote such gifted-ness, see Chua (2010, 81–101).

References

Alkire, Sabina. 2000. The basic dimensions of human flourishing: A comparison of accounts. In *The revival of natural law: Philosophical, theological and ethical responses to the Finnis-Grisez school*, ed. Nigel Biggar and Rufus Black, 73–110. Aldershot: Ashgate.

Aquinas, Thomas. 1968. *On being and essence (De Ente et Essentia)*. 2nd rev ed. Trans. Armand Maurer CSB. Toronto: The Pontifical Institute of Medieval Studies.

Aquinas, Thomas. 1995. *Truth (De Veritate)*, vol. 2. Trans. James V. McGlynn, SJ. Indianapolis: Hackett Publishing.

Aquinas, Thomas. 2001. *Expositio libri Boetii De ebdomadibus (An exposition of the* On the Hebdomads *of Boethius)*. Trans. J. Schultz, and E. Synan. Washington, DC: Catholic University of America Press.

Boland, Vivian, OP. 2007. *St. Thomas Aquinas*. Norfolk: Continuum Books.

Boyle, Joseph. 2002. Free choice, incomparably valuable options and incommensurable categories of good. *American Journal of Jurisprudence* 47: 123–141.

Burrell, David B., CSC. 1990. Creation or emanation: Two paradigms of reason. In *God and creation: An ecumenical symposium*, ed. David B. Burrell, and Bernard McGinn, 27–37. Notre Dame: University of Notre Dame Press.

Caputo, John. 1982. *Heidegger and Aquinas: An essay on overcoming metaphysics*. New York: Fordham University Press.

Carlson, Thomas. 2007. Blindness and the decision to see. In *Counter experiences: Reading Jean-Luc Marion*, ed. Kevin Hart, 153–179. Notre Dame: University of Notre Dame Press.

Cessario, Romanus, OP. 1996. *The elements of moral philosophy.* Washington, DC: Catholic University of America Press.

Chua, Jude Soo Meng. 2000. What limits? Recent positions on the limitation of act by potency in Thomas Aquinas. *Jaarboek 2000 Thomas Instituut te Utrecht,* 42–60.

Chua, Jude Soo Meng. 2006. *A philosophy of education*. Singapore: Prentice Hall.

Chua, Jude Soo Meng. 2008. In praise of folly: Seriously playful curriculum design. *Education Today: The Quarterly Journal of the College of Teachers* 58(4): 18–23.

Chua, Jude Soo Meng. 2009. Saving the teacher's soul: Exorcising the terrors of performativity. *London Review of Education* 7(2): 159–167.

Chua, Jude Soo Meng. 2010. Taking pictures with negative contrast: Edward Schillebeeckx, critical remembrance and policy analysis as practical reason. *Jaarboek 2009 Thomas Instituut te Utrecht,* 81–101.

Clarke, W.Norris. 1994. *Explorations in metaphysics: God-Being-Person*. Notre Dame: University of Notre Dame Press.

Davis, Francis. 2009a. Ed Balls puts the squeeze on faith schools. *The Guardian,* May 1, 2009. Available at: http://www.guardian.co.uk/commentisfree/belief/2009/may/01/ed-balls-faith-schools-funding.

Davis, Francis. 2009b. Gove's plan to set schools free. *The Guardian,* October 9, 2009. Available at: http://www.guardian.co.uk/commentisfree/belief/2009/oct/09/schools-gove-communities.

Davis, Francis and Nathan Koblintz. 2009. *Mutual futures: Ed Balls, Michael Gove and the challenge to faith schools.* Oxford: The Las Casas Institute on Ethics, Governance and Social Justice, Blackfriars Hall, Oxford University. Available at: http://timescolumns.typepad.com/files/mutual-futuresfullfinal.pdf. (Referenced as *Mutual Futures*).

Finnis, John. 1980. *Natural law and natural rights*. Oxford: Oxford University Press.

Finnis, John. 1994. Natural law and legal reasoning. In *Natural law theory: Contemporary essays*, ed. Robert P. George, 146–151. New York: Oxford University Press.

Finnis, John. 1997. Commensuration and public reason. In *Incommensurability, incomparability and practical reason*, ed. Ruth Chang, 215–233. Boston: Harvard University Press.

Horner, Robyn. 2005. *Jean Luc Marion: A theo-logical introduction*. Aldershot: Ashgate.

March, James G. 1978. Bounded rationality, ambiguity and the engineering of choice. *Bell Journal of Economics* 9(2): 587–608.

March, James G. 1991. Exploration and exploitation in organizational learning. *Organization Science* 2(1): 71–87.

March, James G. 1994. *A primer in decision making: How decisions happen*. New York: The Free Press.

March, James G. 1999a. Model bias in social action. In *The pursuit of organizational intelligence*, ed. James G. March, 308–324. Oxford: Wiley-Blackwell.

March, James G. 1999b. Wild ideas: The catechism of heresy. In *The pursuit of organizational intelligence*, ed. James G. March, 225–228. Oxford: Wiley-Blackwell.

Marion, Jean-Luc. 1991. *God without being*. Trans. Thomas A. Carlson. Chicago: University of Chicago Press.

Marion, Jean-Luc. 2002. *Being given*: *Toward a phenomenology of givenness*. Trans. Jeffrey Kosky. Stanford: Stanford University Press.

Miller, Adam. 2008. *Badiou, Marion and St Paul: Immanent grace*. London: Continuum.

O'Rourke, Fran. 1992. *Pseudo-Dionysius and the metaphysics of Aquinas*. Leiden: Brill.

Petrilli, Susan. 2007. Abduction, medical semeiotics and semioethics: Individual and social symptomatology from a semiotic perspective. In *Studies in computational intelligence,* ed. Lorenzo Magnimi and Li Ping, 64, 117–130. Netherlands, Springer.

Rahner, Hugo, SJ. 1963. *Man at play or did you ever practise eutrapelia?* London: Compass Book.

Rivera, Joseph. 2009. The call and the gifted in Christological perspective: A reconsideration of Brian Robinette's critique of Jean Luc Marion. *Heythrop Journal* XLVIII: 1–10.

Robinette, Brian. 2007. A gift to theology? Jean Luc Marion's 'saturated phenomena' in Christological perspective. *Heythrop Journal* XLVIII: 86–108.
Schillebeeckx, Edward, OP. 1969. *God the future of man.* New York: Sheed & Ward.
Schillebeeckx, Edward, OP. 1980a. *Jesus: An experiment in Christology.* New York: Herder & Herder.
Schillebeeckx, Edward, OP. 1980b. *Christ: The experience of Jesus as Lord.* Trans. John Bowden. New York: Crossroad/Seabury Press.
Shanley, Brian, OP. 1996. St Thomas Aquinas, onto-theology and Marion. *The Thomist* 60: 617–625.
Simon, Herbert A. 1997. *Administrative behavior: A study of decision making processes in administrative organizations*, 4th ed. New York: The Free Press.
Wippel, John F. 1984. *Metaphysical themes in Thomas Aquinas.* Washington, DC: Catholic University of America Press.

Education for All and International Cooperation for Education Development: Ongoing Implications for National Policy in the Philippines

Andrew Crow and Thomas O'Donoghue

1 Preamble

Of the 193 member states of UNESCO, the Republics of the Philippines, Djibouti and Angola are the only three that retain a 10-year basic education program (UNESCO 2011a). This is a situation that underpins ongoing disadvantage in a progressively global society where continuing development of frameworks and comparative benchmarks underscore recognition, accreditation and equivalency (Washington Accord 1989; Bologna Accord 1999; TIMSS 2011; PIRLS 2011; PISA 2012). In October 2010, The Department of Education for the Republic of the Philippines released a discussion paper titled "The Enhanced K+12 Basic Education Program" (DepEd 2010). The paper describes a new program for the Department underpinning a national priority initiative for the newly elected Aquino government and represents the next chapter in attempts by successive Philippine governments to reform and enhance the national education system. The K+12 basic education program proposal for reform is set against a domestic backdrop of declining investment in education, declining standards and declining participation rates and student outcomes in both elementary and secondary education (Mapa 2009). Conversely, the regional international environment within which the Philippines exists and interacts illustrates increased investment in education, improving student attainment and a progressive widening of the gap in educational performance between these neighboring economies and the Philippines where comparatives exist (Caoli-Rodriguez 2008; Unicef 2009; SEAMEO-INNOTECH 2003; SEAMEO 2008). Additionally, economic development within the region and continued and strengthening ties between countries under the auspices of the ASEAN and APEC intergovernmental frameworks is placing further pressure regionally on domestic education systems to

---

A. Crow (✉) • T. O'Donoghue
Graduate School of Education, The University of Western Australia,
Crawley, WA, Australia
e-mail: andrew.crow@pearson.com

T.B. Mooney and M. Nowacki (eds.), *Aquinas, Education and the East*,
Sophia Studies in Cross-cultural Philosophy of Traditions and Cultures 4,
DOI 10.1007/978-94-007-5261-0_11, © Springer Science+Business Media Dordrecht 2013

deliver policy positions and outcomes in support of developing labour markets (World Bank 1993; Hirosato 2001; ASEAN 2009). This has direct implications for ongoing attempts at harmonization between systems across the region (particularly in terms of recognition of skills, portability of certifications and knowledge sharing in support of ongoing comparative benchmarking).

This chapter examines the recent history of international cooperation in support of regional education development with particular reference to the UNESCO program, Education for All, and explores the situation that the Philippines finds itself in as it attempts to improve the quality of its basic education system whilst simultaneously bringing it into line with regional and international counterparts in terms of a full 12 years schooling. The chapter is presented in four parts. Firstly a view of the Philippines from the perspective of its national participation in regional programs of cooperation is examined. Secondly a closer examination of education reform programs undertaken in the Philippines since 1990 in conjunction with the international Education for All initiative is presented. Thirdly the current situation and context in terms of proposed and ongoing reform activity is examined and then finally the implications for continuing policy development are discussed. The Education for All program is a specific initiative grounded in the United Nations Universal Declaration of Human Rights (art. 26), which in turn conceives of education as a fundamental human right. As such, access to education and the reform initiatives discussed in the context of the Philippines in this chapter draw upon broader conceptions of social justice, of a form St Thomas would recognize. When communicated to the general public, the language of education policy makers is routinely translated to emphasize the Catholic and universal commitment to recognizing the inherent dignity of the human person.

2 Regional Cooperation: A Philippine Perspective

In the post World War II era of international cooperation and establishment of institutional frameworks and capabilities, the Republic of the Philippines has been both a founding party and continuing contributor. The Philippines was a signatory to the founding of the United Nations in 1945, a signatory to the establishment of UNESCO in 1946 and has participated actively over four separate periods (1947–1950, 1953–1973, 1980–1992, 1995–2000) within the Commission on Human Rights since its establishment to elaborate the post World War II international agenda on human rights (CHR 2011; UN 2011; UNESCO 2011b). In its initial period with the Commission, the Philippines contributed to what has come to be generally regarded as one of the most important documents of the twentieth century, the Universal Declaration of Human Rights that was delivered by the Commission and proclaimed by the United Nations on the 10th December 1948 (UN 1948). During the 28 years it took following the Declaration for the International Bill of Rights to come into force in international law, Article 26 of the Universal Declaration of Human Rights stood as a standard bearer having profound and continuing influence over the development of operating principles and agreed charters emerging from cooperative endeavors in education in the South East Asian region (De Baets 2009).

One such endeavor was the South East Asian Ministers of Education Organization (SEAMEO), which was established as a regional intergovernmental body to foster collaboration on education, science, and culture, following a meeting in Bangkok of Ministers in 1965 (SEAMEO 2011a). Two years before the Association of South East Asian Nations (ASEAN) became a reality and 3 years before the first SEAMEO Charter on education was signed, representatives from key South East Asian countries (Thailand, Laos, Malaysia, Singapore, Indonesia, Vietnam and the Philippines) set about establishing regional centers for cooperation in areas such as Education in Science and Mathematics, Languages, Tropical Medicine, Public Health and Community Nutrition amongst others (SEAMEO 2011b). Similarly, in 1967 the Foreign Ministers of five nations, Indonesia, Malaysia, Singapore, Thailand and the Philippines signed the Bangkok Declaration and brought ASEAN into reality (ASEAN 2011). The Declaration provided a framework for enduring peace, sustainable development and economic cooperation and embodied a commitment to multilateral sharing and the continuing development of education (ASEAN 2008). Likewise, with increasing levels of international engagement, the third annual SEAMEO meeting finally ratified the SEAMEO Charter on June 28, 1968 (SEAMEO 2011a). This set a framework for mutual collaboration between the countries of South East Asia to advance knowledge and mutual respect, to establish joint programs for the development of education, to maintain, increase and diffuse knowledge and to assist in articulating education to the economic and social goals of member states. Fundamentally important to the ongoing development and support for SEAMEO was recognition that regional activities may indeed benefit other countries in the world and the original Charter specifically calls this out as a desired goal. In 1973 the Charter was specifically amended to permit the inclusion of non-regional members as 'associates' to participate in the activities and directions of SEAMEO (SEAMEO 2011b). Not surprisingly with this development Australia and New Zealand joined as regional associates and they were joined by France perhaps with motivations of support for the former French colonies of Indochina given the timing was the closing stages of the Vietnam War. Other European colonial powers that had previously been present in the region (Germany, the Netherlands and Spain) also became involved in supporting education development and reform through the auspices of SEAMEO over the ensuing 15–20 years around the time of the establishment of the forum for Asia Pacific Economic Cooperation (APEC).

For the Government of the Philippines, APEC began as an informal 12 nation Ministerial economic dialogue in 1989 that by 1993 had evolved into a formal pan-regional process that by 1998 would involve 21 nations (APEC 2011). The motivation for APEC at the end of the 1980s is as relevant today as it was then that in a rapidly globalizing world the provision of liberalized trading capabilities within the region for both products and services strengthens the capacity of the region to grow and develop. This has enormous implications for the Philippines in terms of national development given that over half of all world trade and 60% of global GDP is generated within APEC countries and the Philippines has been a part of this dialogue since inception. The Bogor Declaration in 1994 clearly articulates a vision for the region involving cooperative economic solutions, supporting open multi-lateral trade in goods and services, reduction in barriers to trade enabling free flow of capital between economies, strengthening of education and training and knowledge transfer linkages for mutual benefit (APEC 2010). These factors have enduring implications for policies supporting national economic development, education and beyond the region.

The World Conference on Education for All held in Jomtien, Thailand in 1990 set a bold new direction for education. The Jomtien Declaration, signed by 155 countries and 150 organizations, set the path for all children, youth and adults to "benefit from educational opportunities designed to meet their basic learning needs." (UNESCO 1990). From the perspective of timing, the Declaration on Education for All and the establishment within the region of APEC was perhaps coincidental. But the awareness of the Declaration in relation to personalizing learning to the needs, culture and circumstances of the learner, combined with the regional and global economic position of nation states for increasingly transportable skilled labor in support of new and emerging markets, provided a powerful incentive to achieve the goals and realization of the Universal Declaration of Human Rights some 45 years earlier.

Throughout the post World War II period, the Republic of the Philippines has also maintained strong links with funding agencies in support of programs for development. These relationships include those with the International Bank for Reconstruction and Development from 1945 (which would later become the World Bank) and from the 1960s the Asian Development Bank (World Bank 2011a; ADB 2011). The Asian Development Bank was established with strong involvement from Japan as Japan strengthened financial and economic development ties with countries that had primarily been occupied by the Japanese during World War II (Indonesia, Malaysia, Thailand, South Korea and the Philippines). Both the World Bank and Asian Development Bank funding arrangements have supported numerous programs in infrastructure development and more recently poverty alleviation as well as significant regional programs in basic education reform specifically related to Education for All (NEDA 2007).

In participating with regional counterparts, the Philippines has undergone consistent reviews of education relevance, performance and suggested change roughly every decade since the First World War. These are detailed in the Philippine Human Development Network Report (PHDN 2008) and include such programs as the 1925 Monroe Survey focused on secondary education and employment pathways, the 1930 Prosser Survey focused on vocational education in lower secondary years (particularly 7th grade), the 1949 UNESCO mission survey which reconstituted 7th grade, the Education Act of 1953 that mapped the first seven grades of schooling, the 1960 Swanson Survey, the Presidential Commission to survey Philippine Education that recommended as a high priority moving to an 11-year system, the 1991 Congressional Commission on Education recommending that the schooling program be 7 years of elementary and 5 years of secondary education, and the 2,000 Presidential Commission on Education Reforms proposing pre-baccalaureate changes to bring the Philippines into line with other countries.

Two decades on from the Jomtien Declaration the Philippines consistently engages with a wide range of organizations in education programs that involve participation with such entities as United Nations system agencies (UNESCO, UNDP, Unicef), intergovernmental institutions (SEAMEO, ASEAN, APEC), development banks (World Bank, ADB), foreign government overseas development assistance channels in the form of grants or loans (USAID, AusAID, JODA, Govt of Spain), regional and local NGO's (Gawad Kalinga), philanthropic foundations (Ayala Foundation, Gates Foundation), research institutions (RTI), and civil society involvement both contractually and increasingly through corporate social responsibility programs.

Combining the internal working relationships between various agencies of the Philippine Government and the multi-lateral framework and donor partners creates a complex interplay between programs with a wide range of nation building objectives (poverty alleviation, education and training, economic development). To support the management of this complexity the Department of Education utilizes a program management task force created in 1972, which is operated under the auspices of the Office of the Secretary and called the Education Programs Implementing Task Force (EDPITAF) (DepEd 2011). Program coordination, oversight, performance management and reporting all come under the jurisdiction of EDPITAF.

In the light of the inter-relationships between the Philippine Government and the regional instruments of international cooperation it is clear that over an extended period of time the Government has had significant access to and participation in regional dialogue, full exposure to a wide range of education reform thinking and comparative benchmarking, and a mechanism with which to acquire accessible funds to support educational change programs. This unfettered access to knowledge and resources begs the questions "so what are the Philippines doing with it?" and "How are the resources and knowledge being utilized to secure enduring improvements in educational attainment and student experience?"

3 Education Reform Programs 1990 – Present Day

With the objective of providing universal primary education and significantly reducing illiteracy during the decade 1990–2000, the World Conference on Education for All provided the basis for national governments to attack the problem of inadequacy and deterioration within their education systems and to approach the problem of functional literacy within their populations. The Jomtien Declaration on Education for All combined with the proposed Framework of Action provided the capacity for the setting of goals and targets, defining the principles of action and the priority actions to be taken at national, regional, and global levels. UNESCO's 1996 Mid Decade Assessment of the continuing performance of countries against the Education for All goals highlighted a number of challenges pertinent to the Philippines education system (UNESCO 1996). These challenges are in the process of being addressed by both current and planned policy and program areas and include increasing the pace of adoption of primary education, closing the gender gap, raising the quality and relevance of schooling, strengthening the teaching force and improving the teaching process to improve learning, focusing resources to early childhood care and education, establishing a stronger focus on literacy and non-formal education for youth and adults, allocating further resources and better utilizing those resources. As part of the Mid-Decade Assessment it was also recognized that data allowing purposeful comparisons and analysis of systemic progress towards the Jomtien goals were significantly variable in quality and availability. This became a focal point for the continuing Education for All Program.

The Third Elementary Education Program (TEEP) of the Social Reform Agenda within the Philippines (sponsored by the Japan Bank for International Cooperation and the World Bank in 1997) identified poverty as a significant variable influencing education performance (World Bank 1996). The program team classified over one third of all households in the Philippines and two third of all regional homes as being in poverty. Facilities in basic education in these regions were identified as inadequate with teacher and classroom numbers declining in relation to increasing student populations. National achievement test results in poverty areas showed a marked decline with 43% of test takers passing compared to national expectations of 75%. Completion rates in poverty areas were also down on the national average at 63% and were significantly lower (15%) than comparable neighboring economies.

However, the global, regional and in-country assessments of the first decade of Education for All reflected that in the main improvements had been made but that much was still to be done (Peppler Barry 2000a, b; Mellor 2000). For the Philippines, net enrolments rose particularly in elementary education, gender equity trended towards balanced but significant issues remained in areas such as per capita funding for education, teacher student ratios, availability of resources and retention rates in the upper years of schooling. In support of continuing objectives through the second decade of Education for All, the Philippine Government began explicitly declaring power and direction through the political process that included an increasingly decentralized operating model for education.

Through the 2001 Republic Act 9155, the Department of Education was formally created from the Department of Education, Culture and Sports as a policy response with specific references to the Governance of Basic Education (LawPhil 2001). This provided the Secretary with explicit powers to formulate national education policy, formulate a national basic education plan, promulgate education standards, monitor and assess national learning outcomes, undertake national research, enhance the status, welfare, competence, work and conditions for all staff, and enhance the development of learners through local and/or national programs. Given ongoing separatist conflict in the southern Philippines the legislation accommodates the governance and activity of a Regional Education Secretary within the Autonomous Region of Muslim Mindanao. Additionally, Republic Act 9155 specifically notes the policy to support the establishment and use of an "Alternative Learning System" in conjunction with the formal education system in order to address functional illiteracy through informal and non-formal learning means.

AusAID through a cooperative grant process with the Government of the Philippines provided $53m AUD in 2002 for the Basic Education Assistance for Mindanao program focusing on minority and isolated communities. This program also sought to contribute to peace and development initiatives within the region (AusAID 2011).

Building on Republic Act 9155, Senate Bill 1636 presented to the 14th Philippine Congress and House Bill 630, presented in 2004 to the 15th Philippine Congress sought to formally establish the Alternative Learning System as an adjunct bureau of the Department of Education (Senate 2004). The proposed Alternative Learning System was to comprise a complete and independent learning system able to address the various learning styles and needs of learners providing learners with full discretion over the kind and method of learning (including the use of information technology). On September 14th, 2004 the then President of the Philippines, Gloria Macapagal Arroyo through Executive Order 356 brought the Alternative Learning System into reality and established a parallel education vehicle to service the marginalized, the functionally illiterate and the out of school youth (Supreme 2004). This legislative commitment formed the basis of the Philippines core program to support literacy for all as part of strengthening ongoing national development.

The programs administered through the Alternative Learning System were established along two paths (BALS 2007). Firstly basic literacy and secondly elementary and secondary education aimed at delivering accreditation and equivalency for learners not in the formal education system. Most of the target demographic learners are people living below the poverty line in depressed, disadvantaged or underserved communities and include unemployed or underemployed out of school youths and adults, elementary or secondary school drop-outs and leavers, industry-based workers, housemaids, wives, factory workers and drivers, members of cultural minorities and indigenous peoples, persons with disabilities or physical challenges, and inmates or rebel soldiers (DepEd 2007).

The Bureau of Alternative Learning has established an entirely new curriculum for functional literacy, elementary and secondary schooling as well as materials to support a variety of delivery options (distance learning, face to face, print material, digital material etc.). The Bureau also established accreditation and assessment testing materials and delivery capabilities to enable successful test takers to mainstream in the education system of the country (Higher Education Pathway, Access to Technical and Vocational Skills Training Programs, Access to Government Employment Positions).

Additionally, the Alternative Learning System has opened new ways of delivering teaching and learning through the use of a mobile teaching workforce; establishing regional and community based learning centers, by offering flexibility in non-classroom delivery to both in-home and in-workplace locations, and by supporting the establishment of networks of third party delivery partners in Local Government Units. The online delivery component of the Alternative System, "eSkwela" has been in operation since 2005 with progressive development of learning modules, teaching practice guides, refined assessment items etc. In 2009 the program was awarded an Innovation in ICT in Education Practice Award from UNESCO in relation to the Community Based e-Learning Centers Supporting Out of School Youth and Adults (CICT 2009). Results from Accreditation and Equivalency testing demonstrate average pass grades more than 20% higher than the national average from the formal education system. Qualitative investigations indicate a significantly higher level of student engagement than the formal system and a significantly stronger community orientation towards learning as a meaningful endeavor. The community engagement model has been further extended from 2009 with the formation of Municipal Literary Coordinating Councils. One such Council, the Agoo Municipal Literary Coordinating Council in La Union, delivers the program in 49 villages (UNESCO 2009). The structured literacy and post-literacy programs have significantly reduced rates of illiteracy in the region and the Council was awarded seven national literacy awards and the UNESCO Confucius Literary Award for 2009. Municipal involvement has also brokered relationships with locally operating NGO's bringing additional sources of funding for resources and in Agoo's case the ability to also offer scholarships to successful candidates for further learning opportunities (Agoo 2011).

In support of an agenda of functional literacy for all, the Department of Education in conjunction with the National Economic Development Agency's Social Development Committee developed the Functionally Literate Filipino National Plan of Action that was endorsed in October 2005. Senate Bill 2012 presented to Congress on July 26th 2005 proposed the integration of technology into the curriculum and also that a national board of computer education be created. This was followed by the creation of the national framework plan for the integration of ICT's in basic education and the national Schools First Initiative DepEd both of which support underpinning broader government policy in terms of the 2004–2010 Medium Term Development Plan of the Philippines (DepEd 2005). The ICT in Basic Education program provided wide ranging policy and implementation guidance that evolved during 2007 into the National ICT for Education Strategy and the core platform for technology innovation in twenty first century teaching and learning. Additionally, the Schools First Initiative formed the basis of the reform agenda to lift educational outcomes through improving current performance, strengthen accountability and responsiveness, and enhance management and leadership. Key targets are defined as follows. All children entering grade 1 are ready for school; all children in school able to read by Grade 3; teachers having English and subject proficiency; all students obtaining adequate instruction; increased participation in schooling; and increased school graduation rates.

Also in 2005, the Department of Education adopted the Basic Education System Reform Agenda (BESRA), a $200m USD World Bank funded program focused on five integrated key reform policy areas. These areas are strengthened school based management, improved teaching effectiveness and teacher development, enhanced quality assurance through standards and assessment, improved access and outcomes through alternative learning and institutional culture change within the Department.

4 Current Situation

The regional review of progress against the goals of Education for All set at Jomtien (UNESCO 1990) then reviewed and renewed at Dakar (Peppler Barry 2000a) show changes in character over the 20 year period of review. The program of Education for All initially endorsed at Jomtien was premised on the provision of basic education and support for the learning needs of children, youth and adults and this shifted to incorporate gender equity and education quality in the ensuing decade. Likewise in the early stages of the program the lead was taken by intergovernmental and donor agencies and this has similarly shifted with continuing economic growth to being supported by national governments with continuing assistance from donors. This is particularly noticeable in the South East Asian region in the reclassification of economic status of countries by the Asian Development Bank and the shift in monetary provisioning from grants to loans as the domestic capacity to "repay" has risen. The rise of in-region donors, such as Australia's AusAID, and the diminishing reliance on European or North American funding sources (as a direct consequence of the economic crisis situation of the second half of the first decade of the new millennium) has changed the nature of regional relationships, influence and the sense of regional belonging. In particular, the role of civil society in supporting the implementing aspects of the continuing reform programs has grown significantly, particularly within the Philippines where participation within the political processes of education delivery are also becoming more common.

In framing a vision for changes to the current education system the President of the Philippine's Benigno S. Aquino III has said: "We need to add 2 years to our basic education. Those who can afford pay up to 14 years of schooling before university. Thus, their children are getting into the best universities and the best jobs after graduation. I want at least 12 years for our public school children to give them an even chance of succeeding." (DepEd 2010)

Perspectives on the failure of the Philippine education system to serve its citizens continue to emerge. The 2009 review "When Reforms don't Transform" noted that despite long term and significant effort the problems identified as key to systemic change had indeed not changed in nearly a century of educational change programs. These noted problems are high dropout rates, low pupil performance, poor teacher quality, excessive centralization, irrelevant learning materials, and poor financial resources. Despite reviews of programs and projects highlighting many gains in student achievement continuing problems exist such as; increasing dropout rates at grade six level (28–34%); low (less than 50%) rates of completion of secondary school by students who enter the Philippine system at year 1; and a less than 10% completion rate of senior year schooling by students in the poorest areas of the nation such as the Autonomous Region of Muslim Mindanao. The Philippines education system now compares less favorably to those of Laos and Cambodia in terms of net enrollment and completion rates and lags significantly behind traditional comparable neighbors from a developmental standpoint (Malaysia and Indonesia) where enrolment and completion rates in the high 90% range having made strong progress since the joint embarkation on the journey initially begun as the Charters of ASEAN and SEAMEO almost four decades earlier. Comparisons have been drawn with the American system in terms of performance of students in science and math. It has been highlighted that three quarters of a century ago Filipino children enjoyed a status of near equivalence that today ranks substantially lower with less than 1% of students attaining above 75% in national achievement testing and average performance ranking at a low 45.8%. More recent comparatives in terms of testing in math and science through the TIMSS program highlight that even the best performing specialist schools within the Philippine system barely rank at the average performance levels of the best performing systems in the region (Taiwan, Hong Kong, Singapore) and indeed are still lower than traditional developmental comparators such as Indonesia. The most recent national achievement test data (2009) also show a continuing decline in pass rates from previous years albeit small.

Pressure from labor markets for skilled and emotionally mature workers highlights a significant mismatch between the output of the Philippine education system and the requirements of the market. This comes in the form of jobless figures whereby 80% of unemployed workers come from the sub 34 year old age bracket and of these over 70% are high school graduates. This situation is also obviously impacted by the practical belief that there will be available work for such groups to undertake, but such high percentages underscore more fundamental issues that have been examined through market based studies such as the Philippines Skills Survey (Di Gropello 2010). Results of the survey underline growing employer demand for critical thinking, higher order problem solving, initiative and creativity as core competency requirements. It is interesting to note that these core competencies are illustrated as key components of both the formal curriculum and of the alternative learning system.

5 Implications for Policy

As 2015 and the commitments made in relation to Education for All approach, three policy areas appear consistently evident. These are reaching the unreached, pre- and post-primary education, and a continued focus on the quality of education. The Philippines has a strong record in observing, planning and implementing programs of change but a poor record in achieving set goals. Indeed UNESCO indicators for country performance against Millennium Development Goals highlight the Philippines at risk in those related to Education. Additionally, within niche program areas such as the Alternative Learning System that has targeted specific areas of concern and combined that with people, funding, resources and community engagement the Philippines has seen significant success albeit in small cohorts. But success nonetheless, and this possibly speaks to some of the prevailing criticisms of Philippine Education in terms of community and economic circumstance, situational relevance of the education deliverable and meaningful outcomes for learners once they exit the system.

In supporting the drive towards a globally comparable K+12 education system, the Department of Education has outlined its new vision to move forward in the coming years. This vision is to create a functional basic education system that will produce functional and productive citizens equipped with the essential competencies and skills for both life long learning and employment. Central to this will be long-term clarity on the language to be utilized as the medium of instruction. Successive governments have adopted alternative positions in this area with consequences for curriculum, ongoing teacher development, assessment, student attainment and workplace readiness. Parliamentary debates in recent years (Congress 2008) have reflected concerns over declining English language proficiency following the introduction of programs such as the DepED 1974 bilingual policy of English and Filipino as the medium of instruction that was specifically instigated to develop a bilingual nation. The 1989 repeal of the bilingual policy to officiate Filipino as the medium of instruction was quoted by President Arroyo in 2006 as further contributing to declining language proficiency and performance in international benchmarks. As recently as 2003, Executive Order 210 establishes the current position where Filipino and English are both recognized within the education system as languages of instruction. English as a second language from Grade 1. English as the medium of instruction for English, Math and Science from Grade 3 and English for at least 70% of content delivered within all public and private institutions at the Secondary level. Such positions foster long-term requirements for continuous professional development of teachers in English language learning and teaching. Also the recentness of the Arroyo Government policy positions espoused in Executive Order 210 combined with the subsequent Aquino Government directives in terms of K+12 education require sophisticated and ongoing coordination efforts between national agencies (NEDA, CHED, TESDA, DepED etc.) to ensure efficacy of policy implementation.

In establishing its future position, the Department aligns itself with two approaches. Firstly, to give every student an opportunity to receive quality education based on an enhanced and decongested curriculum that is internationally recognized and comparable, and secondly, to change public perception that secondary school education is just a preparation for college; rather, it should allow one to take advantage of opportunities for gainful career or employment and/or self-employment in a rapidly changing and increasingly globalized environment. Social justice, and access to fundamental goods appropriate to the dignity of the human person, as expressed so strongly within the Thomistic tradition, demand no less.

Given the prior results of programs involving large donor funding instruments where significant inputs have not necessarily translated to significant outputs the Philippines needs to be rigorous in the manner in which EDPITAF handles the oversight role. There exists an opportunity for EDPITAF given the experience of overseas development assistance program management to expand these capabilities to other programmatic areas incorporated into ongoing reforms. Indeed it seems pragmatic that this existing vehicle could act in a significantly stronger manner supporting external and internal stakeholder coordination linking activity at the lowest levels through to major programs of change in support of realizing tangible benefits from the financial, resources and human capital investments being made. With this in mind, it is clear from a comparative investment position that the Philippines is under-investing per capita on education. It has also set aggressive timeline targets for transformative change bringing in the full K+12 proposed curriculum within a 7 year planning horizon. Given that national performance has been declining, retention rates have been declining and gender equity has become more imbalanced over an extended period with significant institutional support in terms of Education for All it remains to be seen whether the Philippines has the capacity to institutionalize the K+12 change within one cohort generation. Planning must be undertaken to accommodate a slower rate of uptake and/or slower rate of success in terms of raising outcomes. Likewise the geographic diversity, ethnic diversity and the fact that the Department of Education needs to deliver everywhere from Metro Manila to the rebel conflict zones of the South (an area of continued and significant focus for the last decade with continuing dismal results) highlights the complexity of delivering reform. Indeed UNESCO in its 2010 study "Education Under Attack" notes the significant and enduring impact of educational attainment in conflict zones and also highlights the opportunities in terms of community engagement, social development and reform, and potential target areas of support to redress systemic challenges and ongoing research and development. These factors all need to be considered within the ongoing policy framework development in support of national ideals.

Whilst the Philippines has embraced a K+12 reform agenda and seeks to quickly modernize and align the Philippines to Education for All, the world has moved on. The World Bank announced early in 2011 through its Education Strategy 2020 document its commitment to move beyond Education for All to Learning for All and with it shifted the international education goalposts (World Bank 2011b). The shift reflects a response to a dynamic world and one which builds upon the momentum gained through the Education for All international agenda. With an objective reflecting "Learning for All, Beyond Schooling", grounded in international comparative benchmarks such as PISA and TIMSS and "System Reform, Beyond Inputs" with focal points in knowledge and skills acquisition for growth, development, and poverty reduction, the new strategy sets a new benchmark for national education systems reform. Transparency, accountability, and the drive for efficacy in terms of learning outcomes will become hallmarks of the new global agenda.

The policy makers within the Department of Education in the Philippines will need to be cognizant of the shift from "know that" to "know how" and "know why" (Mooney and Nowacki 2011, ch. 1), somewhat like the underlying differences between the international benchmarking approaches of TIMSS and PISA, to ensure that they are truly comparable in future assessments of performance and efficacy of the education system. International benchmarking no longer restricts itself to knowledge of facts ("know that"), but is moving in deliberate steps to encompass twenty first century Skills and Competencies. These involve educational practices targeting skills as outcomes ("know how"), as well as the capacity for seeing the interconnections among domains of knowledge and the ability to see their meanings in context ("know why"). Sensitivity to these shifting international comparative benchmarks is one way of incorporating appropriate policies and initiatives, but this requires a new way of thinking about education, one that embraces not just "know that" but also "know how" and "know why", so as to secure the realization of educational goals within a social justice framework.

References

ADB. 2011. Philippine country data. Available at: http://www.adb.org/Philippines/main.asp.

Agoo. 2011. Agoo Municipal Literacy Center. Available at: http://www.agooweb.com/.

APEC. 2010. APEC 2010 Bogor Goals. Available at: http://www.dfat.gov.au/publications/trade/APEC-2010-Bogor-Goals.html.

APEC. 2011. History, Asia Pacific Economic Cooperation. Available at: http://www.apec.org/en/About-Us/About-APEC/History.aspx.

ASEAN. 2008. *The ASEAN Charter*. Jakarta: ASEAN Secretariat. 54 p.

ASEAN. 2009. *ASEAN Socio-Cultural Community Blueprint*. Jakarta: ASEAN Secretariat.

ASEAN. 2011. ASEAN history. Available At: http://www.aseansec.org/20024.htm.

AusAID. 2011. Philippine country data. Available at: http://www.ausaid.gov.au/country/country.cfm?CountryID=31.

BALS. 2007. Bureau of alternative learning internal briefing presentation. Available At: http://www.scribd.com/doc/6946396/Alternative-Learning-System.

Bologna Accord. 1999. European Higher Education Area 2010–2020. Available at: http://www.ehea.info/.

Caoli-Rodriguez, R.B. 2008. Asia and the Pacific, education for all, mid-decade assessment: Insular South-East Asia synthesis report.

CHR. 2011. Office of the High Commissioner for Human Rights. Available at: http://www.ohchr.org/EN/Pages/WelcomePage.aspx.

CICT. 2009. *eSkwela: Community based e-learning centers for out of school youths, adults*. Bangkok: UNESCO.

Congress. 2008. Medium of Instruction: English, Filipino or Both? Available at: http://www.congress.gov.ph/committees/commnews/commnews_det.php?newsid=947.

De Baets, A. 2009. The impact of the Universal Declaration of Human Rights on the study of history. *History and Theory* 48: 20–43.

DepEd. 2005. National framework plan for ICTs in basic education (2005–2010). Harnessing ICTs for quality basic education for all. Available at: http://www.fit-ed.org/downloads/National%20Framework%20Plan%20for%20ICTs%20in%20Basic%20Education.pdf.

DepEd 2007. Bureau of alternative learning, accreditation and equivalency internal presentation. Available At: http://www.deped.gov.ph/cpanel/uploads/issuanceImg/A&E%20write%20up.pdf.

DepEd. 2010. Discussion paper on the enhanced K+12 basic education program. Available at: http://www.deped.gov.ph/cpanel/uploads/issuanceImg/K12new.pdf.

DepEd. 2011. EDPITAF projects. Available at: http://www.deped.gov.ph/edpitaf/.

Di Gropello, E., H. Tan, and P. Tandon. 2010. *Skills for the labor market in the Philippines*. Washington, DC: The World Bank.

Hirosato, Y. 2001. New challenges for education development and cooperation in Asia in the 21st century: Building indigenous capacity for education reforms. *Journal of International Cooperation in Education* 4(2): 1–24.

LawPhil. 2001. Republic Act 9155. Available at: http://www.lawphil.net/statutes/repacts/ra2001/ra_9155_2001.html.

Mapa, M. 2009. Failing our students, the state of our right to education. In *Under the seal of impunity, the state of human rights during the Arroyo Administration*, ed. J.M. Villero and B.D. Larin. Philippine Human Rights Information Center, Issue 9. Philippine Alliance of Human Rights Advocates.

Mellor, W. 2000. *Education for all 2000 assessment, global synthesis*. Paris: UNESCO.

Mooney, T. Brian, and Mark Nowacki. 2011. *Understanding teaching and learning*. Exeter: St Andrews Studies in Philosophy and Public Affairs/Imprint Academic.
NEDA. 2007. Philippine overseas development assistance reporting, National Economic Development Agency. Available at: http://www.neda.gov.ph/progs_prj/16thODA/16th_oda-main.htm.
Peppler Barry, U. 2000a. *World education forum final report*. Paris: UNESCO.
Peppler Barry, U. 2000b. *World education forum education for all status and trends 2000*. Paris: UNESCO.
PHDN. 2008. Philippine human development report 2008/2009, Institutions, Politics and Human Development, Human Development Network. Available at: http://hdn.org.ph/2009/05/21/20082009-philippine-human-development-report-2/.
PIRLS. 2011. TIMSS and PIRLS International Study Center. Available at: http://timss.bc.edu/.
PISA. 2012. OECD program for international student assessment. Available at: http://www.pisa.oecd.org/pages/0,2987,en_32252351_32235731_1_1_1_1_1,00.html.
SEAMEO. 2008. Reaching the unreached: Meeting of South East Asian countries to achieve the education for all goals by 2015: Meeting report.
SEAMEO. 2011a. SEAMEO Charter. Available at: http://www.seameo.org/images/stories/SEAMEO_General/About_SEAMEO/SEAMEO_Charter/charter.pdf.
SEAMEO. 2011b. SEAMEO, history. Available at: http://www.seameo.org/index.php?option=com_content&task=view&id=24&Itemid=34.
SEAMEO-INNOTECH. 2003. Philippines country data. Available at: http://www.seameo-innotech.org/resources/seameo_country/educ_data/philippines/philippines_ibe.htm.
Senate. 2004. Senate of the Philippines. Available at: http://www.senate.gov.ph/.
Supreme. 2004. Supreme Court of the Philippines. Available at: http://elibrary.judiciary.gov.ph/index10.php?doctype=Executive%20Orders&docid=0368d70a693860b49058b311ac513dba455a487d5159a.
TIMSS. 2011. TIMSS and PIRLS International Study Center. Available at: http://timss.bc.edu/.
UN. 1948. The Universal Declaration of Human Rights. Available at: http://www.un.org/en/documents/udhr/index.shtml.
UN. 2011. About the UN. Available at: http://www.un.org/aboutun/history.htm.
UNESCO. 1990. UNESCO, education for all, background documents, World conference on EFA – World Declaration. Available at: http://www.unesco.org/education/efa/ed_for_all/background/jomtien_declaration.shtml.
UNESCO. 1996. UNESCO, education for all, background documents, mid-decade assessment on EFA. Available at: http://www.unesco.org/education/efa/ed_for_all/background/mid_decade_amman.shtml.
UNESCO. 2009. UNESCO, Education Municipal Literacy Coordination, Philippines. Available at: http://portal.unesco.org/education/en/ev.php-URL_ID=59429&URL_DO=DO_TOPIC&URL_SECTION=201.html.
UNESCO. 2011a. Member States – Communities. Available at: http://portal.unesco.org/en/ev.php-URL_ID=11170&URL_DO=DO_TOPIC&URL_SECTION=201.html.
UNESCO. 2011b. History, United Nations Education, Scientific and Cultural Organization. Available at: http://www.unesco.org/new/en/unesco/about-us/who-we-are/history/.
Unicef. 2009. Gender Equality in Education, East Asia and the Pacific. Education for All, Mid-Decade Assessment Progress Note. Available at: http://www.unicef.org/eapro/Gender_progressNote_web.pdf.
Washington Accord. 1989. Available at: http://www.washingtonaccord.org/.
World Bank. 1993. *The East Asian miracle: Economic growth and public policy*. Washington, DC: The World Bank.
World Bank. 1996. The Third Elementary Education Program (TEEP). Available at: http://go.worldbank.org/4VSWL8CZ20.
World Bank. 2011a. Philippine country data. Available at: http://go.worldbank.org/25TQT9W0X0.
World Bank. 2011b. Learning for all: Investing in people's knowledge and skills to promote development. Available at: http://siteresources.worldbank.org/EDUCATION/Resources/ESSU/Education_Strategy_4_12_2011.pdf.

Index

T.B. Mooney and M. Nowacki (eds.), *Aquinas, Education and the East*, Sophia Studies in Cross-cultural Philosophy of Traditions and Cultures 4, DOI 10.1007/978-94-007-5261-0, © Springer Science+Business Media Dordrecht 2013

CPSIA information can be obtained at www.ICGtesting.com
Printed in the USA
BVOW03*0819180714

359658BV00002B/2/P